Repair Work Ethnographies

Ignaz Strebel · Alain Bovet
Philippe Sormani
Editors

Repair Work Ethnographies

Revisiting Breakdown, Relocating Materiality

palgrave
macmillan

Editors
Ignaz Strebel
Institute of Geography and Sustainability
University of Lausanne
Lausanne, Switzerland

Alain Bovet
Institute of Geography and Sustainability
University of Lausanne
Lausanne, Switzerland

and

HEG—Haute école de gestion Arc
University of Applied Sciences and Arts
Western Switzerland
Neuchâtel, Switzerland

Philippe Sormani
STS Lab
University of Lausanne
Lausanne, Switzerland

and

CEMS-IMM
EHESS
Paris, France

ISBN 978-981-13-2109-2 ISBN 978-981-13-2110-8 (eBook)
https://doi.org/10.1007/978-981-13-2110-8

Library of Congress Control Number: 2018951034

Cover image: Matt Artz/unsplash

This Palgrave Macmillan imprint is published by the registered company Springer Nature Singapore Pte Ltd.
The registered company address is: 152 Beach Road, #21-01/04 Gateway East, Singapore 189721, Singapore

Preface

Repair work takes place in society and engages technology. Yet, how it does so rarely gets clarified. Early observations and comments on the nature of repair, provisional repair or absent repair where prominently made by Claude Lévi-Strauss in his essay "Bricolage", which appeared in the *Savage Mind* in 1966, or Alfred Sohn-Rethel in his seminal *The Ideal of the Broken,* published in 1979. Whilst these writings touch upon the themes of the present book in one way or the other, they remain largely philosophical, using repair improvisation and practical aspects of human work as a starting point to rethink our relationship with the material environment. The 1980s and 1990s saw the publication of the first studies where repair is not only of conceptual but of ethnographic interest too, where it features as a situated practice that deserves to be observed and described. As developed in the introduction to this book, these pioneering works include Douglas Harper's *Working Knowledge* (1987), Julian Orr's *Talking about Machines* (1996), Chris Henke's 'The Mechanics of Workplace Order' (2000) and Tim Dant's work on Car Repair (2002, 2003). These ground-breaking studies were followed by a series of repair studies, mainly published in individual articles, written by an increasing number of loosely connected international scholars within the fields of sociology, STS, geography and studies in material culture.

The present book follows up on this body of initiatives. In spring 2014, we brought together a substantial number of Repair Work Ethnographers for a three-day workshop at the Swiss Federal Institute of Technology, ETH Zürich, where we were involved in a two-year research project on Repair, Maintenance and Urban Assemblage. The workshop allowed us first to discover a unique bouquet of empirical studies on the repair worlds of workers and work places: locomotive restorers, mobile phone doctors, plant mechanics, medical staff and anaesthetists, librarians, underground workers, bicycle technicians, hobbyists, plumbers and scientists.

Beyond the obvious richness and diversity of these field studies, the workshop allowed the emergence of what Wittgenstein would have called family resemblances between the different contributions, or common properties of repair work to be recognised in various settings. These properties are in particular the broken thing or an *in fracto* situation as the common field of action, the diagnostic work necessary to analyse the problem, the procedures to find a solution and finally, the production of the repaired thing. This common ground motivated the participants to commit for a second meeting, that took place on one of the seven hills of Rome, in the Villa Maraini where the Swiss Institute is located. The goal was then to work on the initial presentation so as to make chapters of a forthcoming book. As unifying as the end of the first workshop was, the second workshop proved less smooth, yet perhaps more stimulating. From an initial confrontation with unknown field realities, the discussion switched to more conceptual questions: What is repair as a practice? How to distinguish repair from maintenance? What is breakdown? Is repair innovation? A sharing culture? A sustainable practice? Though these discussions brought dissatisfaction to some, others rejoiced in them. Eventually, it turned out to be the most ethnographic moment of the process. Indeed, there was agreement in the group that the potential of the collaboration would be less fully exploited in theorisation and conceptualisation than in sticking to the detailed description of the repair instances analysed in each paper.

Holding the book in hand and since our first gathering, a lot of water has run through Rome and through Zurich, down the well-known Tibris river and down the less well-known Limmat river too. During the four years of working, travelling, writing, reading, reviewing, waiting and editing, major conferences and workshops and a number of books on repair and relative topics have been advertised and published. As a matter of course, it is not easy to work on a manuscript and to continue sticking to a process of seemingly 'slow science', when the theme to be treated is about to be intensely discussed, mediatised and to enter the public sphere. However, what kept the process and project going, despite occasional criticism on our refusal to indulge in theorising repair, are the stories, descriptions of situations, movements and workbenches. The impatience to see a book published might be compared to the nervousness pervading repair situations themselves. Everybody wants to see that thing being dealt with and made available for using. Retrospectively, when looking at and listening to the noises and voices of everyday repair work, publications on the topic of 'repairing the world' look more like a glut than like a river. And this even more as descriptions and discussions of so-called real-world repair situations remain an exception and a rarity. So, we hope that this book will at least fill this gap in the literature.

The meetings and the discussions that the project involved were possible thanks to various institutions. Our sources of funding for the first workshop were the Swiss National Science Foundation through international collaboration support and project funding of Repair, Maintenance and Urban Assemblage (Grant No. 149478) and the ETH Centre for Research on Architecture, Society and the Built Environment. The Swiss National Science Foundation through a conference grant and the Swiss Institute in Rome (ISR) have funded our second workshop. Beyond these general funding possibilities, we have many intellectual debts. In particular, we thank colleagues from ETH Zürich and the Swiss Institute in Rome, who participated in the workshop: Michele Luminati, Michaela Schmidt, Marko Marskamp, Sara Keel, Julio da Cruz Paulos, Jan Silberberger and Guelfo Carbone. In addition, Marc Relieu from Télécom ParisTech participated in both workshops and made valuable comments on presentations and papers.

Also, we would like to thank Douglas Harper for permission to use photographs from his ethnography at Willie's workshop and Trevor Pinch for his encouragement to have the book framed and focused in terms of local breakdown and situated repair.

We are grateful to all contributors for engaging in this collaborative work. The mutual feedback on presentation and first versions of texts has definitely contributed to the quality of individual publications and the collection as a whole. For good collaboration and attention to detail, we thank Josh Pitt senior editor, Sophie Li and other members of their team at Palgrave Macmillan.

Last but not least, we are grateful to the Institute of Geography and Sustainability at the University of Lausanne, for accommodating follow-on research on repairing energy transition infrastructure. The repair topic will continue to engage us in the next few years.

Lausanne, Switzerland

Ignaz Strebel
Alain Bovet
Philippe Sormani

Contents

1 **Introduction: When Things Break Down** 1
Philippe Sormani, Alain Bovet and Ignaz Strebel

Part I Settings

2 **Repair Work as Inquiry and Improvisation: The Curious Case of Medical Practice** 31
Cornelius Schubert

3 **'A Good Enough Fix': Repair and Maintenance in Librarians' Digitization Practice** 61
Moritz F. Fürst

4 **Job Done: What Repair Does to Caretakers, Tenants and Their Flats** 89
Alain Bovet and Ignaz Strebel

Part II Networks

5 **Mobile Phone Repair Knowledge in Downtown Kampala: Local and Trans-Local Circulations** 129
Lara Houston

6 **The Dance of Maintenance and the Dynamics of Urban Assemblages: The Daily (Re)Assemblage of Paris Subway Signs** 161
Jérôme Denis and David Pontille

7 **Inquiring and Experimenting with Urban Ecology: Pragmatist Lessons from Public Bicycle Repair** 187
Martin Tironi

8 **Interruptions, Lunch Talks, and Support Circles: An Ethnography of Collective Repair in Steam Locomotive Restoration** 221
Christophe Lejeune

Part III Politics

9 **Negotiating Repair: The Infrastructural Contexts of Practice and Power** 255
Christopher R. Henke

10 **Inside the Bicycle: Repair Knowledge for All** 283
Tim Dant

11 **Repair for the Masses? Gender and Care Work in the Fixers' Collective** 313
Meg Young and Daniela K. Rosner

Part IV Afterword

12 **Repair as Transition: Time, Materiality, and Hope** 337
Steven J. Jackson

Index 349

Notes on Contributors

Alain Bovet is a Professor of Communication at the University of Applied Sciences and Arts Western Switzerland (HEG-Arc, HES-SO) and a Senior Researcher at the Institute of Geography and Sustainability (IGD) at the University of Lausanne. His teaching and research engage with a variety of communication processes, from interpersonal communication to public controversies. He is currently engaged in a video-analysis of the work of caretakers and technicians in urban housing. His recent publications include *Langage, activités et ordre social: Faire de la sociologie avec Harvey Sacks [Language, Activities and Social Order: Doing Sociology with Harvey Sacks]* (Peter Lang, Bern, 2014), co-edited with Esther González-Martínez and Fabienne Malbois, and *La démocratie et ses gènes. Le génie génétique dans l'espace public suisse (1992–2005) [Democracy and Its Genes. Genetic Engineering in the Swiss Public Sphere (1992–2005)]* (Antipodes, Lausanne, 2013).

Tim Dant retired from a Professorship in Sociology at Lancaster University at the end of 2013. Among his seven books are *Material Culture in the Social World* (Open University Press, 1999), *Materiality and Society* (Open University Press, 2005) and *Television and the Moral Imaginary* (Palgrave, 2012). He has written on social, cultural, and critical theory with a particular focus on 'things' and consumption and continues to write on knowledge, phenomenology, morality, mind, guns, repair, and imagery.

Jérôme Denis is Professor in Sociology at the Centre de Sociologie de l'Innovation (Mines ParisTech). He studies data labor in various organizations and explores maintenance practices, notably in urban settings. He is the co-founder of Scriptopolis, a collective scientific blog about writing practices (www.scriptopolis.fr/en).

Moritz F. Fürst is a Doctoral Fellow in the Faculty of Geosciences and environment at the University of Lausanne, affiliated to the Institute of Geography and Sustainability. Fürst's interests focus on knowledge and technology practices in the design and maintenance of information infrastructures and building technologies, particularly in the context of urban sociotechnical change towards data-driven and sustainable energy futures. He received an M.A. in science and technology studies and a B.A. in media and communication studies from the University of Vienna and has worked in the telecommunications, energy, and creative industries.

Christopher R. Henke is Associate Professor of Sociology and Environmental Studies at Colgate University. He has research and teaching interests in science and technology studies, food and agriculture, and environmental studies. Henke is the author of *Cultivating Science, Harvesting Power: Science and Industrial Agriculture in California* (MIT Press 2008).

Lara Houston is a Visiting Researcher at Goldsmiths, University of London. Her research begins with moments of breakdown, where technologies and technological systems are beginning to unravel. She studies the practices of repair, mending and remediation that respond to malfunction and decay; and also the processes of salvaging and wasting that accompany the demise of systems or machines. From 2016 to 2018, she was Postdoctoral Researcher on the ERC project Citizen Sense with PI Jennifer Gabrys, at Goldsmiths, in the Sociology Department. Prior to that, she worked at Cornell University in the Information Science Department, as a Postdoctoral Researcher on the NSF project

Reclaiming Repair with Steven J. Jackson. She has a Ph.D. from the Centre for Science Studies at Lancaster University, on mobile phone repair practices in downtown Kampala, Uganda.

Steven J. Jackson is Associate Professor of Information Science and Science and Technology Studies at Cornell University, Chair of the Information Science Department, and Dean of William Keeton House. His work centers on questions of ethics and value in contemporary technoscience, with special emphasis on problems of time, collaboration, and breakdown, maintenance, and repair in complex sociotechnical systems. Current empirical projects include work on new computational infrastructures and experiential change in the sciences; collaboration and creativity in interactive and new media arts; and problems of technology and social change in post-colonial computing environments. His work has been supported by groups ranging from the Ford Foundation and World Bank to Intel Research and the U.S. National Science Foundation.

Christophe Lejeune is an Assistant Professor at the Université de Liège. He teaches qualitative methods to bachelor, master, and doctoral students. As a sociologist of technique, he first studied virtual communities such as the *Open Directory Project* and *Debian* (a Linux distribution made by volunteers). His interests then moved to local organizations (known as *Linux User Groups*) gathering people interested in the Free Software movement. His current fieldwork (covering *Repair Cafés* and heritage railways) investigates how people with no professional background in technology collectively repair, maintain, and make possible the operation of highly technological devices.

David Pontille is Researcher in the French National Center for Scientific Research and works at the Centre de Sociologie de l'Innovation (Mines ParisTech). His research focuses on writing practices, scientific authorship and evaluation technologies, the maintenance of urban infrastructures. He is the co-founder of Scriptopolis, a collective scientific blog about writing practices (www.scriptopolis.fr/en).

Daniela K. Rosner is an Assistant Professor of Human Centered Design & Engineering at the University of Washington (UW). Rosner's research deals with the promises and pitfalls of design interventions for both social inquiry and practices of technology development. Through design and fieldwork, she studies (and catalyzes) surprising connections between emerging cultures of digital production (including hobbyist fixer groups and feminist hacker collectives) and broader developments in industry and public life. Her work has been supported by multiple awards from the U.S. National Science Foundation, including an NSF CAREER award. She received a Ph.D. at UC Berkeley's School of Information, an M.S. from the University of Chicago and a B.F.A. from RISD. Before UW, she conducted postdoctoral work at Stanford's Program on Science, Technology and Society (STS).

Cornelius Schubert is a Lecturer for Innovation Studies at the University of Siegen. He specializes in science and technology studies as well as medical and organizational sociology. He has conducted research on human–technology interactions in medicine and global innovation networks in the semiconductor industry. He is currently collaborating with computer scientists in the co-design of an augmented reality application to support cooperation on a neurosurgical ward.

Philippe Sormani is a Sociologist working at the intersection of ethnography, ethnomethodology, and science and technology studies (STS). Among his chief publications are *Respecifying Lab Ethnography* (Routledge 2014), a reflexive ethnography of experimental physics at work, "The Local Configuration of New Research Fields", a *Sociology of the Sciences Yearbook* (Springer 2016), and *Practicing Art/Science: Experiments in an Emerging Field* (Routledge 2018), a collection of case studies of particular arts, sciences, and technologies in the making. He is currently based at the STS Lab of the University of Lausanne, Switzerland, and affiliated at the Centre d'étude des mouvements sociaux-Institut Marcel Mauss (CEMS-IMM) of the School of Advanced Studies in the Social Sciences (EHESS) in Paris, France.

Ignaz Strebel is an Urban Geographer and Senior Scientist within the Institute of Geography and Sustainability, University of Lausanne, Switzerland. He is a former member of the executive committee of the Center for Research on Architecture, Society and the Built

Environment, ETH Zurich. His research focuses on ethnographically studying urban transformation 'in the making' in particular work settings, as diverse as the architectural studio, the jury board meeting, the construction site, the vehicles of cleaning crews, the workshops of gardeners and heating technicians, and the offices of municipal building administration, facility services, and building management. He co-directed (with Susanne Hofer) the film documentation *Building Care: That's Why Our Cities Do Not Fall Apart* (2014). His recent book publications are on the sociotechnicalities of caretaker work in residential buildings (*Hauswartung. Für Bauten und Bewohnerschaft*, Birkhäuser 2015) and (with Jan Silberberger) on the *Architecture Competition* (Routledge 2017).

Martin Tironi is a Sociologist at Pontificia Universidad Católica de Chile (PUC). He has a Master in Sociology from Université Paris-Sorbonne V, a Ph.D. from and was a Post-doctoral Researcher at the Center for Sociology of Innovation, Ecole des Mines de Paris. He is a Researcher and Professor at the School of Design PUC. He is currently Visiting Fellow at the Centre for Invention and Social Process at Goldsmiths. He is involved in a 3-year research project (FONDECYT) titled 'Datafication of urban environments and individuals: analyzing the designs, practices and discourse around the generation of digital data in Chile'. His research interests include urban and digital infrastructure, urban mobility, critical data studies, design anthropology, and maintenance/repair studies. Some of his recent publications are: 'Unpacking a Citizen Self-tracking Device: Smartness and Idiocy in the Accumulation of Cycling Mobility Data' (*Environment and Planning D: Society and Space*, 2017) and 'Cosmopolitical Implications in the Prototyping Process: Ethnographic Design Practice at the National Zoo in Santiago, Chile'.

Meg Young is a Doctoral Candidate in the Information School at the University of Washington, working in the Tech Policy Lab. Young explores the downstream impacts of emerging technologies via fieldwork in Seattle, WA with local tech activists and other stakeholders. She received an M.S. in Information and a B.A. in Cultural Anthropology from the University of Michigan.

List of Figures

Fig. 1.1	'The silo blower' as sketched by the ethnographer (© Sketch: Courtesy of Douglas Harper, originally published in Harper 1987)	9
Fig. 1.2	Willie in his workshop working on the ventilator of the silo blower (© Photograph: Courtesy of Douglas Harper, originally published in Harper 1987)	10
Fig. 1.3	Willie adds weld to the bars of the ventilator that were not heavy enough (© Photograph: Courtesy of Douglas Harper, originally published in Harper 1987)	11
Fig. 4.1	Trial for oneself (© Video still: I. Strebel)	94
Fig. 4.2	Display of the repaired thing to the tenant (© Video still: I. Strebel)	96
Fig. 4.3	Course of inquiry and problem solved (© Video still: I. Strebel)	98
Fig. 4.4	The heating rod is the source of the problem (© Video still: I. Strebel)	101
Fig. 4.5	The caretaker re-installs the chopping board next to the sink (© Video still: I. Strebel)	110
Fig. 4.6	Leaving the flat with the oven unrepaired (© Video still: I. Strebel)	112

Fig. 4.7	Formulating the cause of the problem (© Video still: I. Strebel)	115
Fig. 4.8	Giving a tutorial (© Video still: I. Strebel)	117
Fig. 5.1	A software repair tool assemblage, Jason's workshop (© Photo: L. Houston)	142
Fig. 5.2	An image posted on the GSM Forum thread 'Nokia 2600 restart & off', by the forum member *parvez__foysal* (© Photo: L. Houston)	146
Fig. 5.3	An image posted on the GSM Forum thread 'Nokia 2600 restart & off', by the forum member mahbubalfa (© Photo: L. Houston)	147
Fig. 5.4	A screenshot of the front page of the NHL (© Photo: L. Houston)	153
Fig. 5.5	A screenshot of the NHL N91 display solution screenshot (© Photo: L. Houston)	154
Fig. 6.1	The replacement of a flawed signboard (© Photo: J. Denis and D. Pontille)	163
Fig. 6.2	Slides out the PVC sheet (© Photo: J. Denis and D. Pontille)	171
Fig. 6.3	Fitting the sign into the lightbox (© Photo: J. Denis and D. Pontille)	172
Fig. 6.4	Shining the sign (© Photo: J. Denis and D. Pontille)	174
Fig. 6.5	Removing screws (© Photo: J. Denis and D. Pontille)	178
Fig. 7.1	An agent in the process of looking at his bicycle maintenance schedule while a user is going to take a Vélib' (© Photo: M. Tironi)	194
Fig. 7.2	Agent repairing a puncture (© Photo: M. Tironi)	200
Fig. 7.3	A tyre lever in the docking terminal (© Photo: M. Tironi)	200
Fig. 7.4	Agent talking to a user (© Photo: M. Tironi)	202
Fig. 7.5	High school students at a Vélib' station 9th district in Paris (© Photo: M. Tironi)	206
Fig. 7.6	Bicycles with a puncture and bent frame (© Photo: M. Tironi)	209
Fig. 8.1	Operating the car jack (© Photo: C. Lejeune)	227
Fig. 8.2	Arranging the extended wrench (© Photo: C. Lejeune)	230
Fig. 8.3	Inspecting the blower (© Photo: C. Lejeune)	238
Fig. 8.4	Lifting the locomotive (© Photo: Serge Poncé)	244

Fig. 8.5	Engineer's blue (© Photo: Serge Poncé)	245
Fig. 9.1	How to Top and Load Sugar Beets (University of California 1944)	271
Fig. 10.1	Floorworker showing the balls in the bearing race (© Video still: RJ The Bike Guy, available at https://www.youtube.com/watch?v=IrzlHjhckSU)	297
Fig. 10.2	Dirty Biker's diagram and rotating gesture (© Video still: C. Hoyle, available at https://www.youtube.com/watch?v=LNzN9oTQA8s)	299
Fig. 10.3	Dirty Biker showing cone and bearings (© Video still: C. Hoyle, available at https://www.youtube.com/watch?v=LNzN9oTQA8s)	300
Fig. 10.4	Professional showing the cup (Online video: available at https://www.youtube.com/watch?v=yHvpzlcKBKg)	301
Fig. 10.5	Music Maker using cone wrenches (Online video: available at https://www.youtube.com/watch?v=wWEoGm1XY7U)	303
Fig. 10.6	Music Maker replacing balls (Online video: available at https://www.youtube.com/watch?v=wWEoGm1XY7U)	304
Fig. 11.1	Sophie explaining the disassembly of the straightening iron to the fixer volunteer (© Photo: M. Young)	315
Fig. 11.2	Northeast Seattle Tool Library Fixers' Collective (© Photo: M. Young)	317
Fig. 11.3	Workshop space of the West Seattle Tool Library (© Photo: M. Young)	320
Fig. 11.4	Tape measure branded as 'Pretty Tools, Tools Designed with a Woman in Mind' (© Photo: M. Young)	323
Fig. 11.5	As she lingered, she stood physically apart from the rest of the fixers, leaning into observe from a few feet away (© Photo: M. Young)	324
Fig. 11.6	Kay consults Guy (left) and Alex (right) (© Photo: M. Young)	329

1

Introduction: When Things Break Down

Philippe Sormani, Alain Bovet and Ignaz Strebel

Airplanes and artwork, bicycles and buildings, cars and computers, photocopiers and printers, the list could be indefinitely extended and reassembled. Yet, what all its items have in common, as with virtually any commodity, infrastructure or technical artifact, is their occasional breakdown, if not inbuilt obsolescence. To problems of the latter kind,

P. Sormani (✉)
STS Lab, University of Lausanne, Lausanne, Switzerland
e-mail: philippe.sormani@unil.ch

P. Sormani
CEMS-IMM, EHESS, Paris, France

A. Bovet
HEG—Haute école de gestion Arc, University of Applied Sciences and Arts Western Switzerland, Neuchâtel, Switzerland
e-mail: alain.bovet@he-arc.ch

I. Strebel· A. Bovet
Institute of Geography and Sustainability, University of Lausanne, Lausanne, Switzerland
e-mail: ignaz.strebel@unil.ch

I. Strebel et al. (eds.), *Repair Work Ethnographies*,
https://doi.org/10.1007/978-981-13-2110-8_1

sustainable development in and beyond product design, a renewed emphasis on care and craft, as well as various 'do it yourself' initiatives, have become key responses—in short, a genuine *Kultur der Reparatur* (Heckl 2013). Upon closer inspection, this "culture of repair" may, of course, not prove as reliable, revolutionary or self-reliant as its confident promotion and first impression suggest. Neither does it seem exempt from artisanal nostalgia, a gendered division of labor, or otherwise conservative politics. Despite or precisely because of these resonances, a renewed engagement with repair and maintenance fuels many ecological alternatives to current frames of innovation, dominant modes of production, consumption, and circulation, as well as recent efforts to rethink our "broken world" more broadly (cf. Jackson 2014). Drawing upon a multifaceted collection of repair work ethnographies, this book homes in on repair work as an everyday practice of distinctive interest to Science and Technology Studies (STS). The book examines its politics, work settings, and networks in and across a wide range of situations, lay and professional, involving a plethora of objects, big and small, and that in the most contrasting circumstances, up north and down south. In so doing, the book fills an important gap in the literature of STS, whilst making a critical contribution to *relocating* the field's materialist outlook (for a related move, see Jarzabkowski and Pinch 2013). The remainder of this introduction is made of three sections. In the first section, the stated gap in the STS literature—the "still-neglected situation" (Quéré 1998) of repair work—and the heuristic interest of considering material disruptions will be briefly considered. In the second section, we shall briefly reconsider how this heuristic potential has been exploited in some 'classic' ethnographies of repair work. The final section outlines how the present collection makes a difference in the current field of STS, if not social inquiry, more broadly.

Investigating Material Disruption and Repair Work in Situ

Bringing together a wide range of case studies, this book holds together in terms of one leitmotif, the leitmotif encapsulated in the book's title: *Repair Work Ethnographies: Revisiting Breakdown, Relocating Materiality.*

The book's pivotal move, then, is to bring to bear the ethnographic study of repair work as an everyday practice in situ on the key conceptual and theoretical discussions on "materiality," "objectual practice," and "non-human agency" in latter-day STS (e.g., Knorr Cetina 2001; Latour 2000, 2005; Pinch and Swedberg 2008; Quéré 2015; Sayes 2013; Schmidgen 2012). Repair work provides an empirical locus of strategic interest, insofar as its situated practice discloses the primary concerns of participants involved, what happens to be material to them—if only to have a kitchen sink unclogged, your computer debugged, or any other everyday item repaired right away. In turn, recent discussions on "materiality" in STS have remained surprisingly aloof from such everyday concerns (e.g., Barad 2003), not least of which their concrete articulation in situ and actual ordering 'in real time,' as entailed by any ordinary course of repair work (for a related argument, see Ingold 2007). To stick to the prime example, Actor-Network Theory (ANT) remains an ambivalent case in point. On the one hand, it has continuously emphasized the "variable ontology" of its cast of characters (e.g., Callon 1990; Law 2009). On the other, it has tended to frame its empirical investigations in terms of theoretical definitions a priori (regarding "social order" to begin with, Latour 2005), instead of investigating the framing exercise and its local contingency, as part of its empirical domain in particular situations (if only to avoid the "ontological monism" of a one-frame-fits-all approach, cf. Collin 2011).[1] Consequently,

> [W]hat ANT [arguably] fails to do is to study closely the interaction or the lived relationship between human beings and material objects. The empirical work is by and large lacking the detail and precision of the more traditional social studies of technology and many of the textual productions and interpretations are those of the sociologist rather than the actors. It is noticeable that there are very few accounts of the perceptual or tactile interaction between humans and objects in the network, few detailed field observations, photographs or use of video to study the process of the network that would allow the material objects to have a presence in the accounts. What are found in the published studies, are textual forms that are produced sometimes by the human participants – engineers' reports, publicity statements, transcripts of discussions, summatory

> diagrams – but often by the sociologist. These can be excitingly irreverent, entertainingly laden with irony and wit and full of interesting conceptual moves – but these textual devices keep the sociologist in control of the play of interpretations and keep the reader at a safe distance from the lived workings of the network. (Dant 2005: 81)

Therefore, the book at hand focuses on the everyday practice of repair work as its core phenomenon. The book's contributions make explicit the technical expertise and practical reasoning, lay and professional, displayed in actual courses of repair work, as well as the entangled background of material conditions, procedural knowledge, and social circumstances that such work may disclose in situ and in vivo. The "lived workings of the network" thus are to be examined as they take shape in and across particular situations of repair work. How do these situations actually unfold? How far are they extended? What kind of actors, materials, and technologies do they involve, if not co-constitute? Is their extension discussed? And how does the "status of the object" (Pels et al. 2002) *under repair* get re-examined, renegotiated and/or re-established? These are some of the key questions that the book and its unique collection of repair work ethnographies will address. Setting aside their specificities for now, it is worthwhile to point out their common strategy—the empirical focus on material disruptions of everyday routines—and intellectual origins of that methodological strategy and its heuristic potential.[2]

Indeed, a number of philosophers and social scientists have shown the insight that can be gained from *material disruption*. As Martin Heidegger suggested via his phenomenological allegory of the 'broken hammer,' it is only when an object fails to lend itself to its common use that it becomes a concern, that we start thinking about it, regretting the lack of its "'readiness-to-hand' as its normative identity" (Lynch et al. 1983: 224). What is more, the "context of equipment is 'lit up' and 'the world announces itself' when something is unready-to-hand" (Heidegger 1962: 105; in Dant 2005: 88). Material disruption, then, not only problematizes a broken instrument, tool, or object but also the social situation that its ordinary use configures—in short, its "world of everyday life" (Schutz 1973). A close idea runs through *pragmatist philosophy* and John Dewey's theory of inquiry in particular. For pragmatist philosophy, characteristically, it is the unanticipated interruption of established routines that renders indeterminate, if not doubtful, a hitherto unquestioned situation. In turn, this indetermination and doubtfulness may lead participants to start inquiring the situation, grasp its changed character, and thereby reestablish their agency. Indeed, Dewey defines inquiry as an oriented, transformative process:

> Inquiry is the controlled or directed transformation of an indeterminate situation into one that is so determinate in its constituent distinctions and relations as to convert the elements of the original situation into a unified whole. (Dewey 1991a: 108)

Two implied ideas are worth pulling out. First, as Dewey explains, the original situation presents itself as an indeterminate or "open" one, insofar as it is not clear how its components hang together as various aspects of that same situation. To return to our first example: does the clogged kitchen sink have something to do with our eating habits? Second, the original situation, if indeterminate, is not to be considered problematic per se. Indeed, for Dewey, the problematization of the situation hinges upon the initiation of an inquiry. Hence the saying: "a problem well-stated is a problem half-solved" (in Dewey 1991a: 131). As a father, one of the authors thus may ask his daughter: "you didn't finish your plate, what did you do with it?" In turn, as analysts, we may attend to such mundane situations, if only to tease out the instructive character of material disruptions and their eventual repair. Lacking such attention, we miss an opportunity to both learn from and care for, if not transform our social and physical environment (we shall broach Dewey's political theory below).[3]

The attention to the heuristics of material disruption has also played an important role in sociology. In particular, Harold Garfinkel's (1967, 2002) *ethnomethodological program* articulates a phenomenological and pragmatist sensibility. Taking its cue from Heidegger's 'broken hammer,' Garfinkel's work demonstrates the heuristic interest of breakdown and breaching. Indeed, his "breaching experiments" were initial attempts to make problematic and thereby observable the taken-for-granted common-sense knowledge and use of social structure. The main lesson that can be drawn from Garfinkel's and his students' multiple experiments, ranging from devising mock-up psychology experiments to asking strange questions at home, is the following: the socio-material world, for it to be upheld in its normal appearance, requires constant work, including typically "seen but unnoticed" repair and maintenance operations in situ. Conversely, such repair and maintenance operations, being internally related to breakdown and failure, create special opportunities to notice unexpected or hitherto unseen aspects of the so-disclosed world (e.g., its "lived workings" as a networked reality). Additionally, Garfinkel's ethnomethodology both prolonged and subverted Dewey's pragmatism by turning both "inquiry" and "theory" into empirical phenomena, instead of confining them to philosophical discussion.[4]

A few more remarks on Dewey's pragmatism and its implications for political theory are in order to highlight the *political relevance* of repair. As already mentioned, Dewey acknowledges the heuristic potential of crisis, failure and breakdown. In his terms, the confrontation with negative consequences of a transaction leads to the pursuit of an inquiry on the cause, conditions and possible remedies of a problematic situation. In *The Public and Its Problems*, Dewey (1991b) applies his theory of inquiry to modern democracy: when faced with a problematic situation, those who feel indirectly concerned may start a collective inquiry in order to prevent its negative consequences. Dewey considers that the political public is constituted, if it happens to be constituted at all, in and through such a public process of inquiry. Although Dewey's emphasis on indirect concern has often been overlooked, it remains central to understanding why any potentially problematic situation does or does not get politicized. When faced with a problem, one typically attempts to solve it first by oneself, then through institutions or specialists, such as repair workers. It is only when directly concerned participants prove unable to handle the problem that it can become an indirect concern to many and thus possibly become a political issue.[5]

"Classic" Ethnographies of Repair Work

How has this multifaceted interest in material disruption—tied to situated inquiry, social order, and possible politicization—been taken up and developed in the ethnographic study of repair work as an everyday practice? Let us take a brief look at three 'classic' ethnographies of repair work: *Working Knowledge* by Douglas Harper (1987), *Talking about Machines* by Julian Orr (1996), and Chris Henke's article "The Mechanics of Workplace Order" (2000). Common to these studies is an aspiration to present and discuss instances of repair work *in action* and to do that on the basis of ethnographic evidence gathered *in situ*.[6]

Working Knowledge by Douglas Harper (1987) is an ethnographic study of repair and maintenance in "Willie's shop," a repair workshop located in Upstate New York where suburban American culture at the time was practically absent. Willie, the protagonist and privileged informant of Harper's study, is a local mechanic and tinkerer who repairs cars, agricultural machines and other devices, all used by the members of his rural community. For Harper, Willie embodies *bricolage* at its best, what Lévi-Strauss (1966) called a "science of the concrete." Accordingly, Harper describes Willie's work as a mode of thinking in immediate dialogue with what he sees, manipulates, and diagnoses. Second, Harper shows how Willie's work and life are embedded in a world and environment that is continuously shaped by previous projects (materials, tools, various tricks and techniques, accumulated from job to job). Finally, Harper also considers how Willie talks about himself, regarding his life choices in particular. Overall, Harper's ethnography leads to a nostalgic appreciation of repair work and of Willie's skills in particular. Skills and work practices are suggested to disappear with Willie as a person and craftsman. Indeed, *Working Knowledge* has predominantly been read as a study of social change that pits modern and industrial forms of work against its preindustrial equivalents. Consequently, cultivating bricolage skills appears as a lost battle, in a modern society and work environment that continuously suppress it. However, a less evolutionary, more subtle reading shows that Harper recognizes transformation not only *of* work but also *in* repair as innovative practice (as both Graham and Thrift 2007 and Sennett

2008 would do two decades later). A second point on which we want to elaborate is Willie's sensibility to repair as work that assists community members in solving material troubles. This second point resonates with the theme of material disruption, as gleaned from the philosophical and sociological traditions discussed above. How does Harper examine it as an everyday phenomenon?[7]

To begin with, Harper uses a dense set of vignettes, gathered during interviews with Willie. Through the method of 'photo elicitation interviews,' Harper invites Willie to discuss and elaborate on aspects of his work that are captured in the ethnographer's photographs. Harper also uses field notes complemented by hand-drawn sketches of machines and devices repaired by Willie (see Fig. 1.1), thereby enabling the reader to understand the operational logic of his work. When looking at Willie's work, Harper discovers that the choices made during work are not "cultural," "economic" or "artistic" in nature, but are related to the local setting and material world in which this work takes place, and contributes to shaping. Willie appears as a craftsman who is indispensable to those who depend on old machinery, dwellings and vehicles that are often assembled from leftover materials. Willie learned from his father, a blacksmith. His working knowledge is deeply rooted in knowledge of the materials and of how they operate (see also Dewey 1991a: 145). His knowledge of how smoke circulates in a wood stove is a good example of this (Harper 1987: 73).

What interests us here is how Harper elaborates on Willie's skills as he deals with broken things and how to fix them. Harper explores Willie's skills as a 'fixer' with a vignette, which shows how Willie repairs a blower, an agricultural machine that is used to blow silage into a silo (Fig. 1.1). Though the blower has been given up on as completely broken, Willie manages to fix it (Fig. 1.2) by investing work time and patience to recognize that somebody has repaired it incorrectly in the past, and with a technique to balance the bars of the ventilator by adding weld to those bars that were not heavy enough. In so doing, he makes it run very smoothly with no vibration (Fig. 1.3). Willie's knowledge of how the blower should run is clearly a product of the troubled situation and of the resources that he is able to make use of. In that, Willie himself exploits the heuristic potential of material disruption, as explored by Heidegger, Dewey, and Garfinkel.

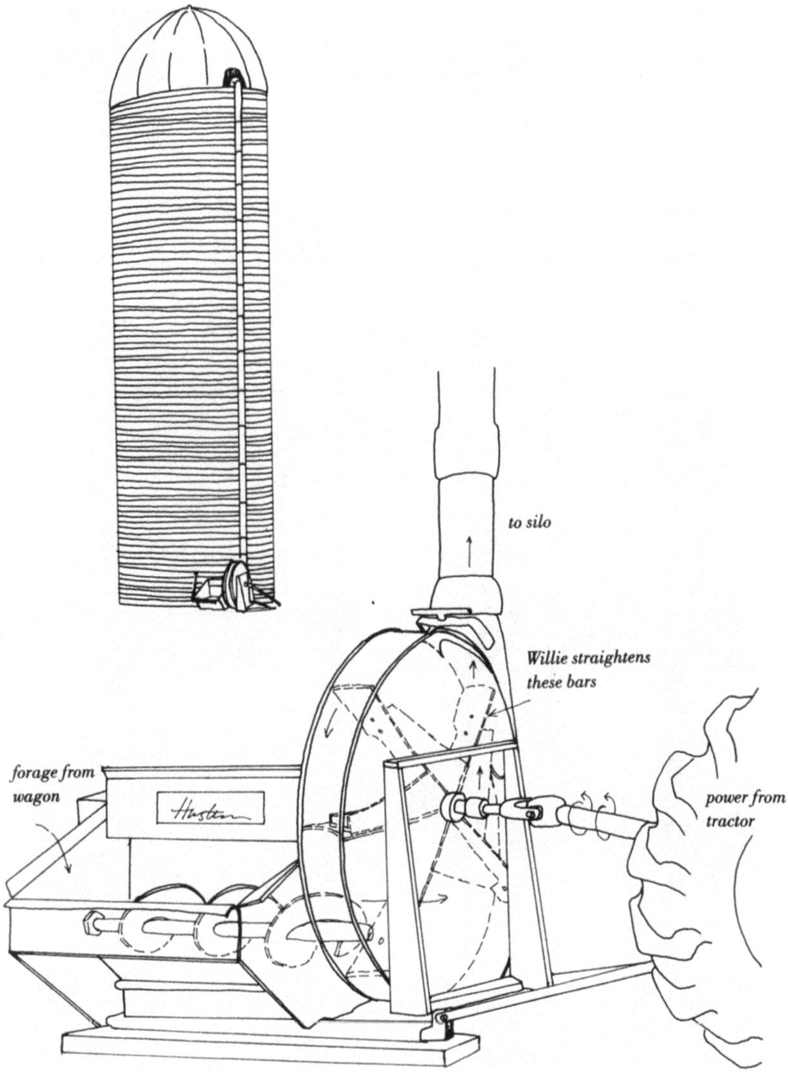

Fig. 1.1 'The silo blower' as sketched by the ethnographer (© Sketch: Courtesy of Douglas Harper, originally published in Harper 1987)

Fig. 1.2 Willie in his workshop working on the ventilator of the silo blower (© Photograph: Courtesy of Douglas Harper, originally published in Harper 1987)

In contrast to Harper's rural ethnography, Julian Orr's (1996) study of photocopier technicians at work is set against the background of a distinctly modern and suburban environment. As its title *Talking about Machines* indicates, the study pays particular attention to how repair work entails both material and conversational engagement, whilst being conducted in a business setting. How do photocopier technicians organize and talk about their work as part of it? How do they conduct their work and, in so doing, have to juggle the peculiarities of the customer, the employer, and the machine—what Orr calls the "Geography of the Service Triangle" (ibid.: 62 onwards)? How do the technicians form their professional identity through their ways of navigating the service triangle and talking about machines? To answer these questions, Orr's study offers an in-depth ethnography of the everyday organization of the professional life of mobile repair workers, an ethnography that emphasizes the situated character of their skills: "actions, or practices, must be understood with reference to the situation of their doing" (ibid.: 11).

Indeed, actions carried out on broken photocopiers cannot be reduced to work plans and protocols that repair should follow. Drawing upon Lucy Suchman's (1987) notion of *situated action*, Orr documents the disparities between work instructions (respectively what technicians are told to do) and the ways workers accomplish work tasks in local contexts (what technicians actually do). Paper jams offer a telling

Fig. 1.3 Willie adds weld to the bars of the ventilator that were not heavy enough (© Photograph: Courtesy of Douglas Harper, originally published in Harper 1987)

example in this respect (Orr 1996: 23–24). Banal as they may appear, they at times require a technician to engage in a lengthy inquiry to locate the origin of the trouble. The technician then draws upon stories told by colleagues, prior experience, customer complaints, locally available tools and ad hoc instrument testing to identify the trouble source. At the same time, he remains keenly aware of the fact that these resources are rarely available as they should be, and this fact shapes the way he proceeds. For Orr, the diagnosis of broken machines aims at "a full representation of the situation" (1996: 121), even if the technician lives in a world where full representation can never be achieved. This observation echoes Dewey's ideal of a terminated inquiry, representing the investigated situation as a "unified whole," whilst offering an invitation to study the technician's locally deployed "ethno-methods" of photocopier repair in detail.[8]

Accordingly, Orr describes, for instance, how a photocopier technician may start to manipulate the broken machine, thereby carrying out a reinterpretation or modification of the initially encountered situation. "Dissolution," then, may become the result of his practice: the problem is not resolved, but actually disappears during "preliminary testing procedures" or through transformations that move "problems out of existence, rather than into a form susceptible of resolution" (1996: 123, note 1). What we see here is that the broken thing does not only trigger repair work, but that the course of this work is likely to modify the problem itself. The relation between the broken thing and the work carried out happens to be constantly changing and, with it, the very sense of "material disruption," ranging from total breakdown to inconsequential occurrence.[9]

Two decades ago, Chris Henke published the article entitled 'The Mechanics of Workplace Order: Toward a Sociology of Repair' (Henke 2000). The subtitle of this seminal article indicates its programmatic ambition beyond previous ethnographies of repair work, including those discussed so far. Indeed, Henke advocates an "expanded analysis of workplace order" and casts his article as a "hopeful first step toward a far-reaching sociology of repair" (2000: 76). To make this first step, Henke's article opts for an eclectic approach, combining Garfinkel's ethnomethodology, emerging STS, and symbolic interactionist intuitions (alluding to the arc of investigation suggested by Dewey's theory of inquiry).

From Garfinkel, Henke takes the idea that "repair" is first of all a practice of *mending order* which, whenever it involves material action, also—crucially—involves 'fixing' social order: "When a workplace breaks down, 'people-repair' goes hand-in-hand with the repair of things" (ibid.: 56). Studying the work of physical plant building mechanics, Henke demonstrates how repair and maintenance restore and normalize workplace environments when they are undermined and made visible through breakdown and failure. If breakdowns initially produce a sense of 'not normal,' this initial sensation engenders various kinds of preliminary activities, including diagnosis and negotiation of technological versus local understandings, and all that even before any repair work is done (if not to avoid it from being done at all, as we have also seen in Orr's ethnography). Conversely, the "sensation of normalcy" (ibid.: 56) in the workplace, for Henke, is a result of complex interactions between the bodies of the workers and their work settings. A case in point is a worker's use of a measuring instrument as an extension of his body to "'look' into ductwork and 'tell' if there is airflow" (ibid.: 64).[10]

However, as Henke points out, "what is needed is a combination of ethnomethodology's strengths (the interactional emergence of order) along with a focus on the material aspects of order" (ibid.: 59). Henke traces this thematic focus on the "material aspects of order" to emerging STS, which notably "explored the embedding of people and things within networks of practical activity" (ibid.: 59–60). Indeed, "when people do work, they are moving through, relying on, and modifying networks of people, ideas, and material artifacts. It is the smooth negotiation of these networks that allows workers to do their jobs" (ibid.: 55). For Henke, then, "workplace order is a 'negotiated order,' shaped through interaction" (ibid.: 61). Building caretakers, for instance, have to negotiate with the users of particular rooms over access, observations made, and previous problems in the building. In the process, caretakers' usually 'invisible work' becomes manifest, "backstage becomes front stage" (ibid.: 73). As such visibilization may lead to "uneasy tension" (ibid.) between users and mechanics, the latter's normalizing strategies may include the display of "oppositional identities," "irony" or "valorizing repair as hard work," with a mechanic claiming to be "the true master of this work setting" (ibid.: 74). However, due to the claimed insufficiency of ethnomethodology, and conversation analysis in particular, Henke's study does not investigate such "agent-client talk" in the face of material disruption in much more detail (cf. Sacks 1995: 391–395).[11]

A New Collection of Repair Work Ethnographies

What difference does this book make? In a nutshell, the book addresses what may be indeed best termed *repair work's "still-neglected situation"* (Quéré 1998; emphasis added). Bringing together a wide range of repair work ethnographies, the book homes in on an instructive array of unfolding situations of routine work, their material disruption, and more or less extended courses of technical inquiry and eventual repair, some of which become "politicized," others not. In other words, the book relocates STS's materialist outlook notably by taking on board Dewey's observation that "one cannot decline to have a situation for that is equivalent to having no experience, not even one of disagreement [and certainly not one of disruption]" (Dewey 1991a, in Quéré 1998: 225). That is to say, if material disruption can only occur from within "complicated social situations" (Jarzabkowski and Pinch 2013: 584), such disruption leads to situated inquiry only, insofar as its parties may extend (or indeed neglect) the boundaries of the situation, both temporally and spatially, but cannot leave it altogether (to wit, "there is no time-out from social order," Garfinkel 2002).[12]

In the light of this internal relation between unfolding situation, material disruption, and technical inquiry, two research interests may be spelled out. *First*, what is of material import to participants, in and as part of their everyday practice, is to be studied in situ. How do participants themselves determine such import, as they go about their everyday business? And how is that everyday business disclosed in its ordinary articulation when it gets materially disrupted? *Second*, participants' inquiries become a phenomenon to be examined in situ, too. How far are these inquiries extended or, conversely, preempted? How do they contribute to articulating, if they do, the unfolding situation and pending repair, if not its potential politicization? And, in so doing, why and to what extent do participants' inquiries recast established configurations of "technological and cultural expertise" (Lindtner 2015: 854)? In line with these two focal research interests, the chapters of this book consider turn in turn the "settings," "networks," and "politics" of repair.

Part I of the volume is entitled *Settings* and brings together a first set of chapters inspired by pragmatism, ethnomethodology and conversation analysis. Accordingly, these chapters all offer in-depth case studies that devote special attention to the detailed development of repair interventions, the practical inquiries they rely on, and the meaning that the participants were progressively able to make and make intelligible to each other, thus progressively configuring what proves material to them, why, just how and in what order. Drawing upon workplace ethnography and video-based studies of interaction, the chapters analyze particular encounters, interventions, and conversations, all of which involve repair and maintenance work in situ.

Cornelius Schubert draws on ethnographic observations of hospital settings to highlight the importance of equipment repair in medical practice. Routinely faced with equipment breakdown and failure, nurses and doctors constantly engage in situated inquiries into the causes, nature and consequences of material problems, and improvise more or less robust remedies to them, thereby taking responsibility for vital issues. Drawing on pragmatist philosophy, Schubert highlights the constitutive dimensions of inquiry and improvisation in repair, notably with respect to how priorities are established in medical practice, between what is judged of material import and has to be urgently addressed, on the one hand, and what is judged of minor significance and can be postponed, on the other.

Moritz Fürst examines how librarians at the Austrian National Library deal with the unavoidable decay of archived and to-be-archived books. The librarians' professional repair and maintenance work is specified by the complex nature of the books and the 'double bind' of their routine treatment. At the same time, they are to be preserved as jewels of cultural heritage and national identity *and* to be made available to interested visitors. Fürst focuses his ethnography on the ongoing digitization process that discloses a tension between preservation and availability. By 'shadowing' librarians at work, Fürst is able to provide the reader with a detailed sense of their distinctive care and professional vision, as well as their local inquiries into book properties, material and digital.

In their chapter, **Alain Bovet and Ignaz Strebel** observe caretakers at work in Swiss residential buildings. They do so as video ethnographers and offer analytic insight into the local repair of damaged objects. In examining selected video clips, they focus on particular closures of caretaker interventions. This empirical focus allows them to qualify caretaker interventions in terms of the distinctive entanglements of social, technical, and material dimensions that the interventions themselves disclose and determine. When it comes to repairing a kitchen tap or an oven that plays a central role in a tenant's life, the caretaker does not only take into account the broken thing, its lacking "readiness-to-hand," but also the irremediably modified viability of the repaired flat. Hence, repair work displays multiple temporalities and distinctive entanglements in situ.

Part II homes in on *Networks* and on how these bring into play new actors, thereby extending the unfolding situation of repair work. Complex connections to wider networks and assemblages of human and material and/or technical entities thus come into view. Truly, the everyday practice of repair work is typically oriented to institutional aspects beyond its local setting, as the contributions to Part I will already make clear. Yet, the networked character of this orientation, as well as its visibilization and extension, all hinge upon the problem at hand and the local resources to have it tackled. The four contributions in Part II examine situations where problems prove too tricky and local resources not sufficient to have them solved, so that actors must and do extend their "situation of inquiry" and establish new connections, beyond any given local context, in view of its eventual resolution and/or relevant repair. The gathered contributions examine not only 'mobile repairables' such as bikes, signpost systems and ancient locomotives, but also the mobility of the resources engaged in repair work, as that mobility proves to be enhanced through digital connection and communication. To use Dant's felicitous expression, these studies come closest to examining the "lived workings of the network."

Lara Houston's chapter is based upon sustained ethnographic fieldwork in downtown Kampala, Uganda's capital, in order to examine the mobilization of multiple resources in and for the repair of mobile phones. Due to the constant evolution of mobile phone technology, local repair shops do and have to rely upon various online networks that provide and circulate the kind of knowledge required to achieve swift and satisfactory phone repair. Through detailed descriptions of actual courses of repair work, Houston shows how local practices get entangled with global knowledge networks. Thereby, she both extends the observed situation, as mobile phone repairers do, and recasts the notion of situated action.

Jérôme Denis and **David Pontille** offer an ethnographic description of maintenance work on Paris' subway signs. Rarely noticed, this professional practice plays an essential role in keeping the subway network intelligible to millions of daily users. Through fine observations of maintenance agents at work, the authors show how they care for the visibility of signs to subway users, the sustainability of the repaired sign, and the accountability of the repair to the subway station superintendent who requested it. The ethnographic approach provides the authors with a renewed and empirically informed perspective on urban assemblage and ANT more broadly.

Martin Tironi describes the continuous maintenance of another Parisian transportation network, the public bicycle scheme called 'Velib.' Tironi highlights the innovative aspect of this network's maintenance. Indeed, the agents engage in socio-technical inquiries on various aspects of the network (users, bikes, locations, time schedules, etc.). In some cases, they actively transform them and, in so doing, the Velib' maintenance agents contribute to urban innovation. As Denis and Pontille's ethnography does, Tironi's study highlights parts of the often invisible and largely ignored work through which urban mobility is sustained. Moreover, Tironi's study highlights the close connection between maintenance routine, local inquiry and occasional reinvention, working out in fine detail the heuristic interest of Dewey's pragmatist notion of inquiry.

Christophe Lejeune addresses public mobility through the study of a peculiar 'old boy network': a Belgian steam engine restoration circle. Through participant observation of the circle's activities, Lejeune provides a vivid account of the extended network of resources that must be mobilized to restore ancient locomotives. He shows that repair relies on the presence of other members for deliberation, advice, and assistance. In addition to that, repair hinges upon a large array of technical and material resources, as made available thanks to members' former employment in various railway companies, as well as their sustained use of digital communication. Such extended networks, when mobilized through the circle's collaboration, make possible the restoration of locomotives and the enhancement of their instrumental and aesthetic qualities (whose gendered production is discussed by Meg Young and Daniela Rosner in the final contribution to this book, see below).

Part III of the volume is entitled *Politics*. That is to say, although we do acknowledge that the networked character of repair work may entail the question "Does repair have politics?," this question is not necessarily made explicit or relevant by repair workers in situ, however extended, networked, and complex their working situation may happen to be (a point that Winner 1999 did not emphasize). Yet, the working situation, of course, does become *politicized* at times, including (among other aspects) the consequences of repair work for concerned, but not directly involved parties. Hence, the contributions in Part III examine such moments of politicization and the politics of repair more broadly. Far from leaving ethnography, they instead use it as a starting point to address political issues of repair work, such as participation, responsibility and consequences on a larger scale, whilst describing how this "larger scale" happens itself to be the product of repair workers' extended inquiries, material interventions, and discursive formulations.

Drawing upon a variety of his own ethnographies, **Christopher Henke** locates repair at the heart of material infrastructure cum social order. Given this centrality, the ethnography of repair topicalizes a "politics of social order," notably by raising issues concerning what should be repaired and what should not, how and by whom, and on which scale, local or institutional. From this conception of repair as a "relational negotiation [...] between people, organizations and institutions, and materiality," Henke draws theoretical and methodological consequences for an ethnography-based sociology of repair, a sociology of repair that takes into account actually unfolding situations and their potentially surprising extensions. In so doing, Henke revisits his earlier position (Henke 2000).

Tim Dant's chapter, in turn, examines repair as a means of empowering users of material objects, in that they are given the means to intervene inside the objects and thus find alternatives to their inbuilt obsolescence and eventual waste. Dant analyzes DIY videos available for free on the internet, videos that instruct their viewers on how to replace worn-out wheel bearings in bicycles. While these videos exhibit a variety of relations to the user and the object, what they have in common is that they invite the viewer to take part in the life of material objects, and thereby take local responsibility for global issues that face us more than ever.

Finally, **Meg Young** and **Daniela Rosner** address a key aspect of this global renewal of our relation to objects, through an ethnographic study of a local "repair culture," a fixers collective in the Seattle region. In that setting, the authors identify a tension between, on the one hand, the empowering trend to take back the reins of the objects that surround us and, on the other, the stubbornness of classical gender roles and scripts, when it comes to doing the actual repair work. Through fine observations of the various obstacles that prevent women from taking an active role (i.e., other than a role as assistant or spectator of the male repairer) in the actual repair work, they highlight some internal tensions of an otherwise progressive and participative "repair culture." In short, they place empirical emphasis on occasioned social discrimination, rather than unlimited network extension.

In his concluding essay, **Steven Jackson** reflects upon common issues of the contributions assembled in the volume. He comments on the diversity of breakdown and repair situations encountered, up north and down south, and on what can be learned from their comparison. As a reminder of and a response to the fallibility of things, the everyday practice of repair provides him with a starting point for renewed theoretical and philosophical reflection. Taking his cue from the revisited "settings," "networks," and "politics," Jackson concludes that repair engages the material and temporal organization of the worlds around us, and thus commits us to an attitude of (reasonable) hope—where "us" includes repair workers, the authors of this introduction, as well as the contributors and intended readers of this volume.

Notes

1. Incidentally, an enlarged notion of "material," emphasizing its physical *and* economical sense, has been proposed to bring economic sociology and STS into closer conversation (see Pinch and Swedberg 2008). Whilst we welcome such a conversation, it still begs the question of how participants themselves determine what is material to them, why now and, indeed, in what sense—the key focus of this volume.
2. Where ANT appears closer to Marxist lines of thought, the present volume is closer to Steven J. Jackson's "broken world thinking" (2014). Jackson draws out the difference as follows: "If Marxism [and, we would argue, ANT] seeks to disrupt the commodity fiction of the object by connecting it backward to moments of origin, discovering the congealed forms of human labor, power and interests that are built into objects at their moment of production, broken world thinking draws our attention around the sociality of objects forward, into the ongoing forms of labor, power, and interest—neither dead nor congealed—that underpin the ongoing survival of things as objects in the world" (Jackson 2014: 230). "Material disruption," in turn, not only questions that ongoing survival of objects, but also invites one to investigate its existential possibility.
3. For recent STS arguments and empirical work on "care in practice," see inter alia Martin et al. (2015), Mol et al. (2010), and Puig de la Bellacasa (2011). In a pragmatist vein, Star and Ruhleder (1996) and Star (1999) had already described a number of key relational characteristics of infrastructure. However, they tended to overlook the fact that infrastructure is maintained in and through both work routines and emergency interventions, and that its ordinary functioning is black-boxed in and through a continuous stream of social and technical practices. Consequently, it is through further pursuing "infrastructural

inversion" (Star 1999) and foregrounding repair and maintenance activities, that we understand (or would understand) what any particular infrastructure is made of (see also Edwards et al. 2009), including presumably any "breaking scientific network" (Margócsy et al. 2017).

4. "Ethnomethodological studies […] have used a variety of methods for breaking the hold of naturalistic sensibility by orienting to 'troubled' instances of activity as an initial condition for revealing order-productive practices" (Lynch et al. 1983: 223). Incidentally, we should note that this kind of breaching strategy has been drawn upon in seminal studies of STS, too (e.g., Latour and Woolgar 1979; Shapin and Schaffer 1985). For systematic discussions of STS in relationship to ethnomethodological inquiry, see Lynch (1993) and, more recently, Lynch (2011a, b) and Sormani et al. (2017).
5. From this perspective, successful repair can be seen as depoliticizing, in that it prevents a problematic situation from becoming a collective political concern (see, e.g., Butler 2015). Conversely, failed repair may be politicizing, in that it leaves the problem unsolved, thus making it available to a political handling. Repair appears then as a crucial phase, maybe even a decisive test, for the political dimension of a problematic situation. While this explains why repair has often been criticized for its normalizing, if not conservative effect (Ureta 2014), it may also work as a reminder of the necessity to find limitations to politicization. Commenting on Hannah Arendt, Claude Lefort (1986) cogently warned that the politicization of every aspect of social life results in totalitarianism. In this view, repair work may be seen as a kind of pre-politicization test for candidate problems.
6. Although they do not exclusively bear on repair and maintenance, these three studies may be considered "classic" ethnographies of repair work, as they are often cited and elaborated upon by more recent studies investigating repair (see, e.g., Dant 2010; Denis et al. 2015; Graham and Thrift 2007). Also it should be noted that repair has not only been studied by STS and social science. Repair is the topic of empirical investigations in work practices and places (see, for example, Gamst 1980), in human–computer interaction (O'Neill 2010), in conversation analysis (Schegloff et al. 1977), in the history of technology (Borg 2007; Corn 2011), in research on commons and collaboration (Baier et al. 2016), in cultural studies (Krebs et al. 2018), in urban

studies (Graham 2010, 2014; Jacobs and Cairns 2011), conservation (Domínguez Rubio 2016; Jones and Yarrow 2013), and in architecture (Langenberg 2018). The difference that this book makes to the existing literature is outlined in Section "A New Collection of Repair Work Ethnographies" below.

7. In asking this question, we suggest to consider repair work as a situated practice, instead of a romantic "escape from modernity, an escape from the externally imposed schedules that impinge on even the briefest moments of experience" (Harper 1987:146). That said, one doesn't necessarily or always exclude the other.
8. In the vein of Suchman's and Orr's work, the *quasi* "paperless office" has become the site and subject of many ethnomethodological studies. For a recent review, see Button et al. (2015).
9. A further instance of this constitutive instability is developed around so-called "war stories" in which technicians narrate their more or less heroic encounters with broken copy machines during their lunch break, where the "use of war stories [becomes] a prominent feature of diagnosis" (125).
10. The initially encountered problems, technical and social, constitute in turn ordinarily occurring "breaching experiments." For a recent discussion in STS of ethnomethodology's interest in various kinds of "trouble," both experimentally induced and ordinarily occurring, see Lynch (2011b).
11. More recently, Tim Dant has investigated repair work under the heading of "material interaction" and paid particular attention to repair workers' skillful manipulation of objects under repair (e.g., Dant 2010; Dant and Bowles 2003). How talk fits into the picture of Dant's approach remains yet to be described (see Bovet and Strebel, this volume). Talk, we should add, is also a major resource for politicizing social relations.
12. The outlined understanding is a far cry from reducing situations to an "interaction order" per se (Goffman 1983), as indeed they may range from a simple greeting exchange (a "minimally complete conversation," Sacks 1995) to Hurricane Katrina's destruction of New Orleans' levees in the aftermath of long-neglected infrastructure or, more recently, Harvey's flooding of Houston. In that sense, if "failure is key" (Graham and Thrift 2007: 7), then only in and as part of a particular situation, ranging from home repair (e.g., Sormani et al. 2015) to "disaster capitalism" at large (cf. Klein 2007).

References

Baier, Andrea, Tom Hansing, Christa Müller, and Karin Werner (eds.). 2016. *Die Welt reparieren: Open Source und Selbermachen als postkapitalistische Praxis*. Bielefeld: Transcript Verlag.

Barad, Karen. 2003. Posthumanist performativity: Toward an understanding of how matter comes to matter. *Signs: Journal of Women in Culture and Society* 28 (3): 801–830.

Borg, Kevin L. 2007. *Auto Mechanics: Technology and Expertise in Twentieth-Century America*. Baltimore: The Johns Hopkins University Press.

Butler, Judith. 2015. *Notes Toward a Performative Theory of Assembly*. Cambridge, MA: Harvard University Press.

Button, Graham, Andrew Crabtree, Marc Rouncefield, and Peter Tolmie. 2015. *Deconstructing Ethnography: Towards a Social Methodology for Ubiquitous Computing and Interactive Systems Design*. Cham: Springer.

Callon, Michel. 1990. Techno-economic networks and irreversibility. *The Sociological Review* 38 (1): 132–161.

Collin, Finn. 2011. *Science Studies as Naturalized Philosophy*. Berlin: Springer.

Corn, Joseph J. 2011. *User Unfriendly: Consumer Struggles with Personal Technologies from Clocks and Sewing Machines to Cars and Computers*. Baltimore: Johns Hopkins University Press.

Dant, Tim. 2005. *Materiality and Society*. Maidenhead: Open University Press.

———. 2010. The work of repair: Gesture, emotion, and sensual knowledge. *Sociological Research Online* 15 (3): 1–7.

Dant, Tim, and David Bowles. 2003. Dealing with dirt: Servicing and repairing cars. *Sociological Research Online* 8.

Denis, Jérôme, David Pontille, and Alessandro Mongili (eds.). 2015. Special issue on maintenance and repair in STS. *Tecnoscienza: Italian Journal of Science & Technology Studies* 6 (2): 5–16.

Dewey, John. 1991a [1938]. Logic: The theory of inquiry (reprinted). In *John Dewey, The Later Works, 1925–1953, Volume 12: 1938*, ed. Jo Ann Boydston. Carbondale and Edwardsville: Southern Illinois University Press.

———. 1991b [1927]. *The Public and Its Problems*. Athens: Swallow Press and Ohio University Press.

Domínguez Rubio, Fernando. 2016. On the discrepancy between objects and things: An ecological approach. *Journal of Material Culture* 21 (1): 59–86.

Edwards, Paul N., Steven J. Jackson, Geoffrey C. Bowker, and Robin Williams. 2009. An agenda for infrastructure studies. *Journal of the Association for Information Systems* 10: 364–374.

Gamst, Frederick C. 1980. *The Hoghead: An Industrial Ethnology of the Locomotive Engineer*. Holt, Rineheart and Winston: New York.

Garfinkel, Harold. 1967. *Studies in Ethnomethodology*. Cambridge: Polity Press.

———. 2002. *Ethnomethodology's Program: Working Out Durkheim's Aphorism*, ed. and intro. Anne Rawls. Lanham, MD: Rowman & Littlefield.

Goffman, Erving. 1983. The interaction order. *American Sociological Review* 48: 854–867.

Graham, Stephen (ed.). 2010. *Disrupted Cities: When Infrastructure Fails*. London: Routledge.

———. 2014. Automated repair and back-up systems. In *Globalization in Practice*, ed. Nigel Thrift, Adam Ticknell, Steve Woolgar, and William H. Rupp. Oxford: Oxford University Press.

Graham, Stephen, and Nigel Thrift. 2007. Out of order: Understanding repair and maintenance. *Theory, Culture and Society* 24: 1–25.

Harper, Douglas. 1987. *Working Knowledge: Skill and Community in a Small Shop*. Berkeley: University of California Press.

Heckl, Wolfgang M. 2013. *Die Kultur der Reparatur*. München: Hanser.

Heidegger, Martin. 1962 [1927]. *Being and Time*. Oxford: Blackwell.

Henke, Christopher R. 2000. The mechanics of workplace order: Toward a sociology of repair. *Berkeley Journal of Sociology* 44: 55–81.

Ingold, Tim. 2007. Materials against materiality. *Archeological Dialogues* 14 (1): 1–16.

Jackson, Steven J. 2014. Rethinking repair. In *Media Technologies: Essays on Communication, Materiality and Society*, ed. Tarleton Gillespie, Pablo Boczkowski, and Kirsten Foot. Cambridge: MIT Press.

Jacobs, Jane, and Stephen Cairns. 2011. Ecologies of dwelling: Maintaining high-rise housing in Singapore. In *The New Companion to the City*, ed. Sophie Watson and Gary Bridge, 79–95. Oxford: Blackwell.

Jarzabkowski, Paula, and Trevor Pinch. 2013. Sociomateriality is the 'New Black': Accomplishing repurposing, reinscripting and repairing in context. *Management* 16: 579–592.

Jones, Siân, and Thomas Yarrow. 2013. Crafting authenticity: An ethnography of conservation practice. *Journal of Material Culture* 18: 3–26.

Klein, Noami. 2007. *The Shock Doctrine: The Rise of Disaster Capitalism*. Toronto: Penguin Random House of Canada.

Knorr Cetina, Karin. 2001. Objectual practice. In *The Practice Turn in Contemporary Theory*, ed. Theodore R. Schatzki, Karin Knorr Cetina, and Eike Von Savigny, 175–188. London: Routledge.

Krebs, Stefan, Gabriele Schabacher, and Heike Weber (eds.). 2018. *Kulturen des Reparierens: Dinge Wissen Praktiken*. Bielefeld: Transcript Verlag.

Langenberg, Silke (ed.). 2018. *Reparatur: Anstiftung zum Denken und Machen*. Berlin and Stuttgart: Hatje Cantz Verlag.

Latour, Bruno. 2000. When things strike back: A possible contribution of 'science studies' to the social sciences. *British Journal of Sociology* 51 (1): 107–123.

Latour, Bruno. 2005. *Reassembling the Social. An Introduction to Actor-Network-Theory*. Oxford: Oxford University Press.

Latour, Bruno, and Steve Woolgar. 1979. *Laboratory Life. The Social Construction of Scientific Facts*. London: Sage.

Law, John. 2009. Actor network theory and material semiotics. In *The New Blackwell Companion to Social Theory*, ed. B.S. Turner, 142–158. Oxford: Blackwell.

Lefort, Claude. 1986. Hannah Arendt et la question du politique. In *Essais sur le politique*, ed. C. Lefort, 64–78. Paris: Le Seuil.

Lévi-Strauss, Claude. 1966. *The Savage Mind*. London: Weidenfeld and Nicolson.

Lindtner, Silvia. 2015. Hacking with Chinese characteristics: The promises of the maker movement against China's manufacturing culture. *Science, Technology, & Human Values* 40 (5): 854–879.

Lynch, Michael. 1993. *Scientific Practice and Ordinary Action: Ethnomethodology and Social Studies of Science*. New York: Cambridge University Press.

Lynch, Michael. 2011a. Ad hoc special section on ethnomethodological studies of science, mathematics, and technical activity: Introduction. *Social Studies of Science* 41 (6): 835–837.

Lynch, Michael. 2011b. Harold Garfinkel (29 October 1917–21 April 2011): A remembrance and reminder. *Social Studies of Science* 41 (6): 927–942.

Lynch, Michael, Eric Livingston, and Harold Garfinkel. 1983. Temporal order in laboratory work. In *Science Observed: Perspectives on the Social Study of Science*, ed. Karin Knorr-Cetina and Michael Mulkay, 205–238. London: Sage.

Margócsy, Daniel, William Rankin, and Sergio Sismondo (eds.). 2017. Special issue: Breaking scientific networks. *Social Studies of Science* 47 (3).

Martin, Aryn, Natasha Myres, and Ana Viseu. 2015. The politics of care in technoscienc. *Social Studies of Science* 45 (5): 625–641.

Mol, Annemarie, Ingunn Moser, and Jeannette Pols (eds.). 2010. *Care in Practice. On Tinkering in Clinics, Homes and Farms*. Bielefeld: Transcript Verlag.

O'Neill, Jacki. 2010. Making and breaking troubleshooting logics: Diagnosis in office settings. In *Ethnographies of Diagnostic Work: Dimensions of Transformative Practice*, ed. Monica Büscher, Dawn Goodwin, and Jesssica Mesman, 35–53. Houndmills: Palgrave Macmillan.

Orr, Julian E. 1996. *Talking About Machines: An Ethnography of a Modern Job*. Ithaca and London: Cornell University Press.

Pels, Dick, Kevin Hetherington, and Frédéric Vandenberghe. 2002. The status of the object: Performances, mediations, and techniques. *Theory, Culture & Society* 19: 1–21.

Pinch, Trevor, and Richard Swedberg (eds.). 2008. *Living in a Material World: Economic Sociology Meets Science and Technology Studies*. Cambridge: MIT Press.

Puig de la Bellacasa, Maria. 2011. Matters of care in technoscience: Assembling neglected things. *Social Studies of Science* 41 (1): 85–106.

Quéré, Louis. 1998. The still—Neglected situation. *Réseaux: The French Journal of Communication* 6: 223–253.

Quéré, Louis. 2015. Retour sur l'agentivité des objets. *Institut Marcel Mauss—CEMS Occasional Paper* 25: 1–12.

Sacks, Harvey. 1995. *Lectures Vol. I & II*. Oxford: Basil Blackwell.

Sayes, Edwin. 2013. Actor-Network Theory and methodology: Just what does it mean to say that nonhumans have agency? *Social Studies of Science* 44: 134–149.

Schegloff, Emanuel A., Gail Jefferson, and Harvey Sacks. 1977. The preference for self-correction in the organization of repair. *Language* 53: 361–382.

Schmidgen, Henning. 2012. The materiality of things? Bruno Latour, Charles Péguy and the history of science. *History of the Human Sciences* 26: 3–28.

Schutz, Alfred. 1973 [1962]. Common-sense and scientific interpretation of human action. In *Collected Papers I: The Problem of Social Reality*, ed. Alfred Schutz, 3–47. The Hague: Martinus Nijhoff.

Sennett, Richard. 2008. *The Craftsman*. New Haven and London: Yale University Press.

Shapin, Steven, and Simon Schaffer. 1985. *Leviathan and the Air-Pump: Hobbes, Boyle, and the Experimental Life*. Princeton, NJ: Princeton University Press.

Sormani, Philippe, Ignaz Strebel, and Alain Bovet. 2015. Reassembling repair: Of maintenance routine, botched jobs, and situated inquiry. *Tecnoscienza: Italian Journal of Science & Technologie Studies* 6: 41–60.

Sormani, Philippe, Morana Alač, Alain Bovet, and Christian Greiffenhagen. 2017. Ethnomethodology, video analysis, and STS. In *The Handbook of Science and Technology Studies*, eds. Ulrike Felt, Rayvon Fouché, Clarke A. Miller, and Laurel Smith-Doerr, 4th ed., 113–137. Cambridge: MIT Press.

Star, Susan Leigh. 1999. The ethnography of infrastructure. *American Behavioral Scientist* 43: 377–391.

Star, Susan Leigh, and Karen Ruhleder. 1996. Steps toward an ecology of infrastructure: Design and access for large information spaces. *Information Systems Research* 7: 111–134.

Suchman, Lucy. 1987. *Plans and Situated Actions: The Problem of Human-Machine Communication*. Cambridge: Cambridge University Press.

Ureta, Sebastián. 2014. Normalizing transantiago: On the challenges (and limits) of repairing infrastructures. *Social Studies of Science* 44: 368–392.

Winner, Langdon. 1999. Do artifacts have politics? In *The Social Shaping of Technology*, ed. Donald MacKenzie and Judy Wajcman, 28–40. Buckingham: Open University Press.

Part I

Settings

2

Repair Work as Inquiry and Improvisation: The Curious Case of Medical Practice

Cornelius Schubert

Introduction

> You have to know a lot to do nothing
>
> —Medical Aphorism

The work of doctors and nurses has to deal routinely with any number of breakdowns. In addition to diagnosing, treating, and caring for 'broken' human bodies, doctors and nurses must ordinarily attend to diverse technical malfunctions and organisational disruptions. The daily routines of modern medical practice are thus always in need of situational adjustments and adaptations, because machines may unexpectedly stop working or relevant information may be missing from the patients file. This chapter will take a closer look at the latter kind of repair in medicine, i.e. the repair activities necessary to maintain the

C. Schubert (✉)
University of Siegen, Siegen, Germany
e-mail: cornelius.schubert@uni-siegen.de

I. Strebel et al. (eds.), *Repair Work Ethnographies*,
https://doi.org/10.1007/978-981-13-2110-8_2

technical and organisational conditions of modern medical practice. It seeks to shed light on these ordinary and often invisible, yet at the same time, fundamental repair practices of medical and nursing work.

I will do so by unfolding the repair work of doctors and nurses in modern medical practice both empirically and conceptually. Empirically, I will draw on ethnographic vignettes from participant observations in surgical operating theatres. Conceptually, I will discuss these observations with questions concerning (a) their specificity and generalisation, (b) the relation of routines and disruptions, as well as (c) the relation of humans and artefacts. My arguments will draw on pragmatist philosophy as well as interactionist sociology in order to position medical repair work within a broader theoretical frame. This frame will be tailored to the specifics of medical practice by elaborating on the ideas of inquiry and improvisation as a combination of routine and disruption as well as the interlocking of humans and artefacts. With this approach, I do not seek to replace or conflate the term *repair* with either *inquiry* or *improvisation*, but hope to draw a more nuanced picture of medical repair work. This is necessary for cross-linking the sociology of repair with the sociology of medicine and the sociology of technology, where repair issues have been a central concern, albeit not necessarily by that name.

The chapter is organised as follows. The ethnographic vignettes portray typical situations of medical repair work. These observations come from routinely scheduled surgical operations in different hospitals which I visited in the early 2000s. At the time, I was working in a research project studying the routines and risks of distributed agency in hospitals. Although my focus was on human–machine interaction and distributed agency, I quickly became aware of the manifold routines, compensations, and improvisations, which make up organised medical practice. The vignettes I have selected for this chapter should provide sufficient glimpses of the field in order to follow the main lines of discussion and to enable the reader to relate my empirical material with that of the other chapters. The vignettes thus simultaneously serve as a base and as an opening for the subsequent elaboration. The reader should also bear in mind that the following sequences of medical repair work are by no means extraordinary, but constitute the perfectly normal situated adjustments of complex work environments. They do not disclose problematic safety issues, but instead highlight the ongoing accomplishment of a stable course of events in the face of frequently occurring practical uncertainties or ambiguities.

The Routines of Breakdown and Repair in Anaesthesia

Breakdown and repair are constitutive features of modern medical practice. But they become particularly crucial during surgery and intensive care, when the vital functions of the patient's body are sustained by a complex array of machines, nurses, and doctors. A breakdown of the life-supporting equipment would in consequence lead to the inacceptable death of a patient. Medical devices are therefore specifically tested for reliability and robustness and intensive care wards and surgical operating rooms have backup devices available on short notice. In case of technical malfunctions or failures, some automated life support systems can also be replaced manually, e.g. in case of artificial respiration. However, disruptions must not necessarily be due to malfunctioning machinery. They also occur whenever the orderly integration of the distributed diagnostic and therapeutic activities becomes fragile, when information is missing from the patient files or ambiguous information is displayed on clinical monitoring systems. These issues will be addressed in the ethnographic vignettes throughout the paper. I will start with a vignette that shows a series of situated adjustments, i.e. mundane and routine repairs, by the anaesthetic team during tumour removal surgery.

<u>Vignette 1: Routine adjustments</u>

> It is half past seven in the morning. Operations typically start at 7.00 AM and the patient is already lying on the bed in the ante-room of the operating theatre with the anaesthetist, the anaesthetic nurse and myself as participant observer. The anaesthetist is preparing peridural anaesthesia (PDA) in order to supplement general anaesthesia and to administer post-operative analgesia after the cancer in the abdomen has been

removed. Inserting the PDA catheter into the epidural space of the spinal cord requires the patient to sit still while the anaesthetist carefully paces the tube in the spinal canal close to the nerve fibres. But the patient, a 76-year-old woman weighing about 60 kg, is very drowsy and cannot sit upright by herself. So, the anaesthetist checks her medical record and finds that she has been routinely sedated on the ward prior to surgery. Yet, the dosage is too high for her age and weight – he refers to himself and me (two men between 30 and 40 weighing around 75 kg) by commenting: "this much would send us to sleep". He says he does not mind a higher dosage of sedatives per se, since it calms the patients, but this clearly interferes with delivering the PDA. He tells me that he has addressed this issue often with his colleagues on the ward, but they do not keep it in mind and continue to give a higher dosage rather than a lower one. In order to commence with the PDA and not to delay the operation, the nurse offers to hold and support the patient and then asks me to assist her by getting an infusion from the cupboard and connecting it.

The anaesthesia then commences without further trouble. About an hour later, at 8.30 AM, the anaesthetist notices that the tube from the antibiotic infusion has become loose and that the fluid is dripping on the floor. He reconnects the tube. He also observes that the patient's body temperature is low and increases the room temperature to counter the cooling effect of the surgical intervention. At 9.00 AM, the pulse rate displayed on the electrocardiogram is lower than the pulse readings taken from the other sources like pulse oximetry. The anaesthetist adjusts the settings of the electrocardiogram and the pulse readings again match those from the other devices.

After the operation is finished around 11.30 AM, the anaesthetist types the anaesthetic protocol into a software program in an adjacent room. While he is typing in the report, the computer freezes and has to be restarted. He is also missing the adhesive patient labels he needs to attach to the paper version of the report. The anaesthetic nurse and a surgeon start to look for the labels, but cannot find them. They eventually turn up under a mat during the following operation.

This vignette demonstrates how doctors and nurses routinely handle breakdowns and organise repair. It also shows the diverse situations out of which the need for adjustments arises. Repair in this sense does not mean that what is broken needs to be taken away to be fixed elsewhere, it also does not entail stopping the course of actions until the breakdown has been mended. Instead, the vignette highlights repair as a basic and necessary feature for maintaining an orderly course of events, a situational re-adjustment of actors and artefacts. This notion of repair work builds on an understanding of order as an ongoing accomplishment emphasised by Garfinkel (1967), where actors routinely complete the incompleteness of contingent situations.

Repair, Inquiry, and Medical Work

Inquiry and Pragmatist Philosophy

I want to elaborate this notion of repair work in medicine by relating the ethnographic vignette to the pragmatist concept of "inquiry" (Dewey 1938) as a process of situated problem-solving. Dewey's basic understanding is that "inquiry grows out of an earlier state of settled adjustment, which, because of disturbance, is indeterminate or problematic" (ibid.: 34). The need for re-adjustment or repair, in other words, emerges out of the previous routines. Inquiry does not pit routine and repair as polar opposites, but highlights that routine and orderly conduct are the fabric of everyday experience out of which indeterminate situations may arise. The drowsiness of the patient in the episode above was the result of an organisational routine on the ward. Organised medical practice harbours countless potentials for such misalignments and uncertainties. In addition to conflicting routines as a source of indeterminate situations, repair practices themselves are grounded in routines. Especially in professional contexts like medicine, they are often taught, learnt, trained, and repeated until they become incorporated into bodily skills. Medical emergency protocols such as resuscitation procedures are a prime example of this. Asking the ethnographer to assist the nurse while she assists the anaesthetist is no formally sanctioned repair practice, but it refers to informal expectations in the operating theatre to provide mutual assistance across formal responsibilities if need be.

The inherent uncertainties of medical practice have been a central tenet of sociological research in medicine for over fifty years now. Parsons, for instance, marked uncertainty as a defining feature of medical practice, in which physicians will always only have limited control over the outcome of a treatment (Parsons 1951: 447). Fox later reconstructed medical school as a "training for uncertainty" (Fox 1957). Despite the scientific and technological advances of modern medicine, uncertainty cannot be fully resolved and new uncertainties may sometimes be created through them, as later ethnographic episodes will show. This insight into the uncertain nature of medical practice resonates well with the pragmatist understanding of practice and experience as potentially uncertain, ambiguous or problematic. In the much-cited words of James: "experience, as we know, has ways of boiling over, and making us correct our present formulas" (1907: 222). Dewey phrases it in a similar vein: "The distinctive characteristic of practical activity, one which is so inherent that it cannot be eliminated, is the uncertainty which attends it. [...] Practical activity deals with individualized and unique situations which are never exactly duplicable and about which, accordingly, no complete assurance is possible" (Dewey 1930 [1929]: 10).

However, such a general notion of uncertainty applies to all practical doings. Dewey therefore distinguishes the indeterminate situation that gives rise to inquiry from "uncertainty at large" (Dewey 1938: 105). Indeterminate situations are not doubtful in and of themselves, but it is a "unique doubtfulness" (ibid.) that defines each situation as a situation constituted by the specific conditions and materials given. Because each indeterminate situation is uniquely structured, the process of inquiry is equally not a general method, but a specific procedure occasioned by the indeterminate situation. In other words, even though uncertainty is an inherent characteristic of practical activities, it is not random, but shaped by the practical activities themselves, because they shape the characteristic traits of a specific indeterminate situation. The same is true for medical inquiry. Inquiry in medical practice is not an abstract questioning, but takes the form of situated "practical judgements", i.e. "judgments of a situation demanding action" (Dewey 1916: 335). Inquiry thus turns the relatively open indeterminate situation into a more directed problematic situation, a situation in which the definition of the problem already is the first step in a specific course towards its resolution. Such problem-solution packages are, again, dependent on the concrete situation.

If we think of medical repair work in terms of inquiry, Dewey reminds us not to start with cognitive reflexion, but with a concrete situation, its material and social fabric as well as its historic evolvement. For instance, repair work in surgery and intensive care is always organised with respect to the well-being of the patient. The operating room with all its equipment as well as the skilled doctors and nurses thus prefigure potential solutions. Repair work as inquiry then transforms indeterminate situations into determinate situations as it restores an orderly course of interactions by actively re-adjusting the components of the situation, replacing some and fixing others. The instances of reconnecting the tube, regulating the temperature or adjusting the settings of the electrocardiographic monitor in the first vignette are part of this continuous practical re-configuration.

Medical Work and Interactionist Sociology

One of the most prominent sociological approaches to medical work as a situated and practical activity comes from interactionism. Strauss and his colleagues developed the concept of "negotiated order" in a study of a psychiatric hospital (Strauss et al. 1963). They were intrigued by the observation that the social order of the hospital is upheld by a plurality of situated negotiations rather than its formal organisation. The cooperative order of the hospital, it can be argued, is in constant need of calibration and adjustment.

Strauss et al. do not use repair as a conceptual term; they make the general claim that norms and rules, i.e. the social order of the hospital itself, are neither fixed nor immutable, but subject to constant review and adaption through collective negotiations. The need to reconstitute social order continuously, ultimately leads to an understanding in which negotiations or repairs cease to be exceptions to a given norm, but rather become its underlying mechanisms. It also emphasises that the collective working order must always be created and maintained through situated adjustments.

As with the case of James' and Dewey's general notion of practical uncertainty, conceiving negotiations on this level of abstraction is too general to make specific claims about medical practice. Strauss et al. therefore hint at constellations, in which negotiated order is most likely to occur. Taking the hospital as a case in point, they assume that the prevalence of negotiated order increases, (a) if different occupations have to cooperate, (b) if the occupations are themselves heterogeneous, and (c) subsequently follow different orientations, and last not least (d) the personnel tends to move between different organisations (Strauss et al. 1963: 168). While Strauss subsequently developed the negotiation framework as broader approach to the questions of social order (Strauss 1978), he and his colleagues continued to study medical practice in hospitals.

Their seminal study on the "Social Organization of Medical Work" (Strauss et al. 1985) again highlights the negotiated order of the hospital and how it must be understood as the intertwining of different kinds of work in which the doctors, nurses and patients are engaged. These kinds of work include machine work, safety work, comfort work, sentimental work, articulation work, and the work of patients. Strauss et al. position their study as a critical analysis of the dysfunctional rationalisation tendencies of medical work in a system that has to cope with an increase of chronic illness. The tensions between rationalisation and the care of chronic patients then become visible in the multitudes of work that are not represented in formal job descriptions and organisational procedures, but that are vital to the successful treatment of an extended illness trajectory. Out of these different kinds of work, machine, safety, and articulation work are the most relevant for this chapter.

"Machine work" (ibid.: 40) is a category that includes the monitoring and maintenance of medical equipment by professional service staff, but also by the doctors and nurses. The authors list numerous strategies and tactics to cope with malfunctioning machinery, like keeping a larger stock of backup machines, having screwdrivers and other tools ready for quick repairs or devising creative work-arounds with sticky tape. Such "makeshift" repairs could be considered as lay and en passant versions of proper repair, but they were nothing less than essential under the given technical and organisational conditions. With the increase of technology in medicine, we can assume, the need for local maintenance and repair increased the same way.

"Safety work" (ibid.: 69) is related to machine work, because care and treatment largely rely on medical equipment. Malfunctioning machines may directly affect patient safety, especially in intensive care medicine. Strauss et al. point out that even though hospitals often employ professional service staff, the sheer amount and heterogeneity of the equipment soon become too difficult to be handled for centralised services. Safety work and machine work thus necessarily include the continuous monitoring, maintenance and repair of machines and their interfaces with the bodies of doctors, nurses and patients.

"Articulation work" (ibid.: 151) is again close to the concept of negotiation, as it denotes the practical coordination of activities as adjustments to the formal organisation of work in a bureaucratic structure. Strauss et al. emphasise that articulation work directly concerns the situated repair of unforeseen and unforeseeable disruptions of medical practice and they list nine potential sources of such disruptions (ibid.: 153): (1) the contingencies of complex treatments, (2) the tension between standardisation of treatment and individual care, (3) the competition of patients for resources, mainly for the staff's time, (4) the contingencies brought in by the patients themselves, (5) the disruptions caused by medical technologies, (6) the impartial integration of necessary supplies and services within the hospital, (7) the interference between different kinds of work, (8) possible differences in the evaluation of work, and (9) possible activities of patients outside the hospital.

These nine sources of disruption are not isolated factors and many disruptions occur where they intersect. However, listing possible disruptions is not an indicator that modern medical practice is continuously on the verge of sliding into chaos. Rather, Strauss et al. accentuate that the stability of the social order in a hospital is not pre-given, but always depends on the situated practices of monitoring, maintain and repairing this order.

In the final chapter of the book, the authors bring their discussion around to larger issues of "mistakes at work" (ibid.: 239). The potential for mistakes is, of course, unlimited. They introduce the term "error work" (ibid.: 242) in order to account for the work of correcting mistakes, inevitable or preventable ones, from professionals or clients, and referring to clinical or organisational issues. Error work essentially embodies the necessary repair of disturbances brought about by the complexity of treatment itself.

This short recapitulation of medical work as machine, safety, articulation, and error work highlights the different faces of repair work in medicine. Repair work can be orientated towards machines, towards organisational routines, towards previous treatments, and towards administrative protocols. What Strauss et al. have shown is that such kinds of repair work do not decrease with rationalisation, but that they are a necessary condition of formal organisation. In a more poignant way of saying: the scientific, technical and administrative formalisation of medicine holds at the same time the potential for disruptions and the need for repair. Timmermans and Berg emphasise this in a study of medical protocols: "Tinkering, having the leeway to adjust the protocol to unforeseen events and repair unworkable prescriptions is a *prerequisite* for the protocol's functioning" (Timmermans and Berg 1997: 293).

As I have tried to argue, pragmatist philosophy and interactionist sociology help to gain a deeper understanding of both the general features and the specific characteristics of medical repair work. Dewey's notion of inquiry makes us sensitive towards the fundamental situatedness of practical problem-solving. Strauss et al. emphasise the different faces of medical repair work: technical, clinical, personal, organisational, etc. Both pragmatist philosophy and interactionist sociology share the general prospect that repair is a mundane and equally essential aspect of practical activity. Repair work in medicine thus mirrors the repair work portrayed in the other chapters of this book as maintenance of socio-material order. In addition, the examples from medical repair work direct our attention towards the en passant modes of repair that accompany the primary medical objective of treating or 'repairing' patients.

Multiplicities of Repair in Medical Work

I would like to round off this discussion by coming back to ethnographic observations with the aim to complement the picture I have drawn so far. Thus, I would like to make explicit and question aspects of medical repair work that have remained implicit until this point. The first aspect is that of the situated order of repair and the second concerns the relation of breakdown and repair.

A focus on the situated order of repair might, in fact, overemphasise the ad hoc emergence of repair procedures. In many cases, however, medical repair work follows simple and practical recipes. Reconnecting tubes, regulating the temperature, or adjusting the settings of the electrocardiographic monitor are routine practices that mostly comply with the operating manual of the machines. Considering repair as practices highlights this recurrent nature of repair procedures. Practical repair inquiries then start with first trying out known solutions, which in many cases will solve the problem. Alternative attempts are only made if the situation remains indeterminate.

Medical emergencies are a case in point. Emergency procedures are formally standardised and trained as part of the job. Even though they are not truly universal, but adapted and adjusted locally (Timmermans and Berg 1997), they rely on routinised completion and swift action. Emergency procedures have to be trained, because they typically are not part of the daily work and I very rarely observed critical situations during my observations. The following vignette is from one of those rare occasions and will show the trained response to a sudden drop in blood pressure, which happened shortly after narcosis was administered.

<u>Vignette 2: Following Emergency Procedures</u>

> The patient is a diabetic woman with a history of cardiac arrhythmia, who is scheduled to get a pacemaker. She did not sleep well the previous night because of pain and is now quite nervous before surgery. She is lying awake in the OR for longer than usual, because the current information on her blood coagulation is missing from the file. After the anaesthetist called the lab and the ward, he only got yesterday's results and decided to check the blood coagulation again – hence the delay. While the team is waiting for the current results, an emergency occurs in a neighbouring OR and people come rushing over to get some equipment from this OR. The ensuing hectic further increases the patient's distress.
>
> After a while, the OR becomes quiet again and the current results come back from the lab. The coagulation is sufficient and surgery can

> proceed, so the anaesthetist administers narcosis according to protocol and without any further disturbances. A few minutes after the intubation, the blood pressure of the patient suddenly drops and the monitoring alarms go off. The anaesthetist rushes to the patient's side and starts cardiopulmonary resuscitation by manually pressing on the patient's chest. He instructs the nurse to prepare a shot of adrenaline. The surgeon rushes by just as quickly and takes over cardiopulmonary resuscitation from the anaesthetist who then administers the adrenaline. After about three minutes, heartbeat and blood pressure are stable again, but after an extended discussion of the situation between anaesthetists and surgeons, the team decides not to operate and to try a different form of narcosis the following day.

This, of course, is a dramatic sequence. It contrasts the small adjustments from the previous vignette with a medical emergency procedure. Yet, both ordinary and emergency repair work follow predefined (formal and informal) patterns of action. Especially, emergency responses leave little room for negotiation. In the case of the second vignette, negotiation happens after the patient is again stable and continues during day, as doctors and nurses keep discussing what happened in the morning. The difference between ordinary and emergency repair work then lies in the temporal extension and the (in-)conspicuousness of the repair situation. Ordinary repair work is seldom talked about. The emergency repair work, which lasted for about two to three minutes, is followed by a detailed assessment of the situation and a careful consideration of possible options. The situated order of repair can thus include many different facets. Medical repair work always combines routine and flexibility in various degrees and as a routine practice, we can indeed expect a large degree of regularly recurring repair work.

This brings me to the second aspect concerning the relation of breakdown and repair. Most emergency situations constitute a clear relation between breakdown and repair, what is broken and how it can be fixed. But, in many ordinary situations, the relation of breakdown and repair is less obvious. Rebooting the computer after the documentation software crashed does not fix the program; rather, it creates suitable workarounds and "ways of going on" (Collins et al. 1997). Such temporary and partial repair work should not be considered inferior to fixing things "properly". Rather, it can be used as an analytic tool to unravel the subtleties of what constitutes a breakdown in medical practice.

Some of the repairs I have observed in the operating theatre are, in fact, not aimed at fixing malfunctioning technology. Instead, they serve to bypass technical functionalities that are likely to disrupt the normal working order. For instance, some infusion pumps used for administering intravenous anaesthesia are equipped with a sensor to detect if the tube connecting the patient is blocked. The tube needs to be placed in the sensor on the front panel of the pump, or else, the machine will not function. However, the sensor is quite sensitive and prone to false alarms. Such pumps can be 'fixed' by placing a short piece of empty tube into the sensor, which effectively bypasses the impractical safety feature. This form of repair depends largely on the experience and skill of the staff in the way that experienced doctors and nurses are more likely to bypass safety features than novices.

If we conceive medical repair work in a broader sense to include such practical bypasses and situated "workarounds" (Gießmann and Schabacher 2014), the identity of the breakdowns shifts from biological disorders and technical malfunctions to organisational misalignments. These breakdowns often emerge at the fault lines between local practices and formal prescriptions, as the personnel seek to maintain a stable order of events by preventing known instances of disruption. Even if this means to put machines and instruments into working order by circumventing their safety features and creatively engaging with impractical technical or organisational conditions. Turning off or disabling alarms does not necessarily imply a safety hazard for the patient; rather, the personnel chooses to rely on other means, i.e. their experience and skill, for making sure that the operation proceeds as planned (Mort et al. 2005; Schubert 2011). In this broader sense, medical repair work includes the activities aimed at maintaining order and keeping things going as they were (Henke 2000).

The ethnographic vignettes and the insights from medical sociology show that repair has many different empirical facets in the context of medical work. Medical practice should not be conceived as a uniform and rationalised application of scientific knowledge; rather, situated medical practices emerge from messy and heterogeneous assemblages of knowledge, technologies, bodies, work, and organisations (Berg and Mol 1998). The empirical studies of medical work show how instances of repairing, bypassing, tinkering, or adjusting the social-technical ensembles of modern medicine constitute one of its defining features. Indeed, as Strauss and his colleagues have stressed, it is at the core of the social organisation of medical work. This in turn leads us back to the question in which ways medical repair work shares some more general features with work and practical activity outside of medicine and in how far medical repair work is defined by the characteristics of modern medical care. I would like to answer this question by addressing the relation of medical repair work with the overarching issues of "inquiry" (Dewey 1938), "work" (Hughes 1951), and "situated actions" (Suchman 2007).

Connecting Medicine and Repair: Inquiry, Work, and Situated Action

Conceiving repair as a form of inquiry links medical repair work with practical activities in general. Dewey emphasised that "scientific inquiry" is not fundamentally different to "common sense inquiry", rather, common sense inquiry occurs "for the sake of settlement of some issue of use and enjoyment, and not, as in scientific inquiry, for its own sake" (Dewey 1938: 61). Medical repair work can be understood as a form of common sense inquiry that settles issues of use in medical settings. It is not inquiry for its own sake, like the scientific inquiry of medical research, but inquiry for the sake of maintaining a stable sequence of treatment. Considering repair as ordinary and common sense inquiry closely connects repair with fundamental issues of practically creating order as an ongoing accomplishment (Garfinkel 1967). Both understandings highlight that medical repair work as I have sketched it out so far is not repair for its own sake (this would be the actual treatment of the patient), but rather repair en passant—as "ways of going on" (Collins et al. 1997).

On the flipside of this, it is difficult to elaborate on the specifics of medical practice by using a general concept such as inquiry. It is equally difficult to use a general notion of uncertainty as the defining feature of medical practice (Parsons 1951: 447). As already said, the distinctive features of medical repair work as inquiry do not emerge from universal uncertainty, but from the "unique doubtfulness" of the given situation (Dewey 1938: 105). This unique doubtfulness, however, is not arbitrary, but characteristic for modern medical practice. We can already find it in the nine sources of disruption by Strauss et al. (1985: 153) listed above. Each single one of them may not be unique to medicine, e.g. possible differences in the evaluation of work, but the way in which they overlap and interfere creates specific lines of tension that distinctly mark modern medical work (despite all its differences emphasised in Berg/Mol 1998) as specific technically mediated, socially organised and sensually embodied practices.

Since the concept of inquiry relates repair to practical activity in general, it is useful to add the notion of repair as "work" (Hughes 1951) in order to sketch out some further distinctions. I use the term repair work rather than repair by itself throughout the paper so as to emphasise two interrelated issues. First, all work necessarily implies some form of repair and second, repair can often be considered some kind of work. In addition, understanding repair as work allows for a comparative perspective that may relate medical work to other types of work along some common themes. One of the common themes addressed by Hughes is that of "routine and emergency", which points to the asymmetry that especially in repair work, "one man's routine of work is made up of the emergencies of other people. [...] Both the physician and the plumber do practice esoteric techniques for the benefit of people in distress" (Hughes 1951: 320). A second theme is that of "mistakes and failures" (ibid.), which highlights the necessary training, skills and knowledge to competently perform certain tasks (cf. Harper 1987). This includes the possibility of delegating failures through the division of labour and the question of defining what counts as failure and what does not. Both themes help to conceive repair work in general and medical repair work in particular as mostly mundane routine work, which deals with reoccurring problems and established solutions.

In addition, they can also serve as orientations for addressing the relations between breakdown and repair in the context of medical work. We find abundant empirical examples for this in the sociological studies of medical work, not only by Strauss and his colleagues, but also in studies of medicine as profession (Freidson 1970), medical knowledge (Wright and Treacher 1982), medicine as skilful practice (Pinch et al. 1996) or managing medical failures (Bosk 2003 [1979]). Again, what makes medical repair work a particular case is not a singular trait, but the combination in which mistakes, failures, and breakdowns relate to skill, competence, and repair.

The third overarching issue next to inquiry and work is that of "situated action" (Suchman 2007). The concept of situated action bears close resemblance to the notions of inquiry and work, because it draws on ethnomethodology, pragmatism and interactionist sociology (ibid.: 67). But in addition to inquiry and work, it specifically puts situated actions in the context of high-tech organisational settings and human-machine interactions by underscoring that "every course of action depends in essential ways on its material and social circumstances" (ibid.: 70). Whereas the original concept of situated action (Suchman 1987) was primarily framed as a critique to the shortcomings of cognitivist and rationalist understandings of action as rule following behaviour, it has been extended towards the mutual constitution of bodies, technologies, and organisations, not least with respect to medicine, as Suchman points out herself: "Over the course of an anaesthesia, agencies involved in the maintenance of vital bodily functions are progressively delegated from the patient as an autonomously embodied entity to an intricately interconnected socio-material assemblage and then back again" (Suchman 2007: 264).

Situated action thus bears on two important issues of repair work. First, it emphasises that repair knowledge is not confined to cognitive aspects, but that it essentially is distributed knowledge, i.e. it is mediated through tools and devices, embodied in skills and practices, organised through the division of labour, documented in manuals and shared through stories (Orr 1996; Henke 2000; Dant 2010). Second, it highlights the socio-material constitution of repair and the importance of considering (technological) breakdowns. Recent repair studies have emphasised the decay and brittleness of materiality and technology, arguing that they are far from being stable self-sufficient entities, but in constant need of repair and maintenance (Jackson 2014; Denis and Pontille 2015). Both issues can also be found in the areas where technology, work, and organisations overlap in situated action research and medical sociology. Workplace Studies in medicine are a prominent example of this (Heath et al. 2003). Actor-Network Theory has been another important influence for understanding the role of technology in medical practice (Timmermans and Berg 2003). These studies show how high-tech medical settings such as telemedicine (Mort et al. 2003) and anaesthesia (Goodwin 2007) essentially depend on constant repair and the situated reconfigurations of knowledge, bodies, and technologies. Last but not least, the prominent concept of "care" in the sociology of clinical settings has been extended towards other fields of study (Mol et al. 2010). Care and repair share many empirical and conceptual overlaps, but I hesitate to see them as synonymous. For once, the term care primarily relates to humans and the term repair to technology. Yet, the care studies and repair studies mentioned above vehemently argue for transgressing this distinction by drawing on practice-based approaches such as situated action and ethnographic research.

The conceptual connections between inquiry, work, and situated action help to frame medical repair work within the questions I have posed in the introduction concerning its specificity and generalisation, the relation of routines and disruptions, and the relation of humans and artefacts. Inquiry, work and situated action can equally be used for zooming into the specifics of medical practice just as they allow for the zooming out from medical repair work to more diverse empirical cases and a broader conceptual discussion. Inquiry, work, and situated action are also sensitive to, if not primarily concerned with, the relation of breakdown and repair. The relations of humans and artefacts are at the centre of situated action, but neither inquiry nor work are exclusively human endeavours. Dewey, for instance, emphasised the role of tools and instruments in the process of inquiry as practical activity. Hughes, in a similar vein, stressed the control over technical contingencies as a defining feature of professional work. However, the empirical multiplicity of repair practices in medicine and beyond, as well as the conceptual plurality for understanding them, makes it at the same time difficult to look for more than conceptual similarities and empirical differences (or vice versa).

Whereas inquiry, work, and situated action have been useful for unlocking the practices of medical repair work on a conceptual level,

I would like to add to this discussion by elaborating a concept of improvisation, which is sensitive to the concrete practices of repair work in modern medicine. Improvisation is a theme in medicine, medical sociology, the sociology of technology, and the sociology of repair alike and thus well suited to bring or hold together the multiplicities of medical repair work. Improvisation should be considered a "sensitising concept" in the original sense of Blumer (1954), i.e. in a strict methodological sense to move "out from the concept to the concrete distinctiveness of the instance instead of embracing the instance in the abstract framework of the concept" (ibid.: 8). As a sensitising concept, improvisation is open to the particularities of medical repair work so that it can be refined through empirical analysis, but it also provides guidance for directing the empirical gaze.

My interest in improvisation as a concept actually developed alongside my ethnographic observations, as I was trying to make sense of all the material I was collecting. Working in the spirit of grounded theory, I was looking to come up with conceptual ideas based on my empirical research. I started by distinguishing the ordinary breakdowns of medical work within a triad of routine, compensation, and improvisation, depending on the amount of repair work needed to maintain working order (Schubert 2006: 128, 2007). Because what I considered improvisation was actually thought of as routine by the doctors and nurses, I adopted the term "routine improvisation" (Schubert 2006: 125) in order to denote its mundane character as well as to underscore the skills and competences of medical improvisation rather than popular connotations of makeshift botching. In short, I did not start out by using improvisation as a sensitising concept, but developed it from the data as a descriptive as well as analytic term or concept (cf. Bruni et al. 2013). I will now discuss improvisation as a sensitising concept by going back to some of my empirical examples and by connecting it with more recent conceptual reflexions.

Medical Repair Work as Improvisation

The third ethnographic vignette serves as an example for the ordinary breakdowns and repairs which cannot be directly traced to single bodies or technologies. It portrays a situated disruption that is subsequently repaired using situated means. In this sense, the situation is unique, but at the same time, the vignette is representative of typical modes of repair in medicine.

Vignette 3: Repairing a Fully Functional Defibrillator

The patient on the operating table is scheduled for a second coronary artery bypass operation. According to the patient's file, an automatic defibrillator was implanted during his first bypass surgery roughly 6 years ago. This device is used to prevent cardiac arrhythmia. It automatically monitors the patient's heart functions and can correct abnormally fast and uncontrolled contractions or even cardiac arrests. Because coronary artery bypass surgery essentially includes cardiac arrest (the vital function are transferred to the cardiopulmonary bypass or 'heart-lung machine'), the defibrillator has to be switched off, lest it interferes by falsely reactivating the heart. In this case, however, the team cannot find any information in the file, if the defibrillator has indeed been deactivated by the cardiologists on the ward prior to surgery or which specific model and make the defibrillator is. It is also unclear, who is responsible for ordering the deactivation.

The surgeon blames the cardiologists on the ward for not doing their job properly, but other team members indicate that it is the responsibility of the surgeon to oversee all the necessary preparations. In any case, pointing the blame does not solve the problem and the anaesthetist calls the ward in order to get the information, but is unable to contact any of the cardiologists in charge. The surgeon demands that a cardiologist comes over from the ward to deactivate the defibrillator and that all activities are stopped until this happens. The anaesthetist calls the ward again on the phone, but again to no avail. Now the surgeon declares that he will go to the ward personally in order to clear the situation and get hold a cardiologist.

After the surgeon has left, further phone calls are made in which the possibility is discussed to deactivate the defibrillator by placing a magnet on it from the outside. After a short while, a magnet is acquired (from where I do not know), yet nobody wants to take the responsibility for this non-standard procedure. This is when a senior surgeon enters the operating room, who, after the situation was explained, confirms that a magnet actually does solve the problem and that he has done so himself in a previous operation. But still no one wants to take the responsibility and the anaesthetist declares that if a decision is not reached within 30 minutes, he will start to wake the patient and stop the surgery altogether. After having left the operating theatre for a couple of minutes, the surgeon who confirmed that a magnet does work comes back and says that he will take responsibility. He asks for the magnet, which was put in a sterile glove in the meantime, and places it on the chest close to the defibrillator. After all personnel are back in the operating theatre, the surgery commences.

The breakdown in this vignette arises because the procedure of coronary artery bypass surgery essentially redefines the functionality of the defibrillator as problematic. Yet, the repair work undertaken is not oriented towards the device at first. What causes the problematic situation is that the status of the device (active/deactivated) is unclear. The defibrillator should have been disabled on the ward, but it might have been forgotten and there is no information in the file. The first attempts at repairing the situation are then levelled at the missing information. The file is checked again and the anaesthetist contacts the ward by telephone. Searching for a solution quickly coincides with question of responsibility for the breakdown. But this issue also remains unclear in the division of labour between surgeons and cardiologists. The status of the device remains obscure and the attempts at clarification turn out to be unsuccessful. We can see that the initial efforts at repairing the situation are oriented towards maintaining formal procedures, i.e. retrieving missing information and identifying official responsibilities. It seems that the team is somewhat uncertain how to proceed after their initial attempts fail. Subsequent telephone calls result in the information that a magnet can be used to deactivate the defibrillator from the outside. But the personnel in the operating room hesitate to proceed as nobody wants to take the responsibility for this informal workaround. The magnet is then only used after a senior surgeon (who is not assigned to this operation) personally takes responsibility.[1]

Such situated re-purposings of available means are an inherent feature of medical practice and they also make up a central element of improvisation. Therefore, it comes as no surprise to find that the discussion about improvisation is nothing new in medicine itself, where it is emphasised as an artistic skill, which, for instance, enables medical personnel to deliver services in emergency situations such as war or accidents (Cubasch 1884). This understanding of improvisation clearly separates two modes of medical practice: ordinary and emergency medicine. Improvisation is related to the latter mode in which the creative mis- or re-use of medical means is warranted if not indispensable under the conditions of scarce material and temporal resources. Cubasch (ibid.: VI) condenses this down to the advice not to lose your wisdom along with your instruments, because following prefabricated rules in wartime medicine is even less feasible than in routines of peacetime work.

Improvisation and Situated Action

We find improvisation once more connected to skill and competence, but also to the situational inability of doing something properly. It is not so much a situation of breakdown and repair as of scarcity and re-purposing. We could indeed presume that the need for improvisation subsides under normal working conditions. This is, however, not the case. First, because all work, even in highly automated settings such as the iron and steel industry, always contains a "chance for improvisation" (Popitz et al. 1957: 64), i.e. the necessary adaptations needed to maintain a stable course of events by resolving the tension between variations in the workflow and the need for preventing errors. In addition to material resources, time can also become a scarce resource. Many industrial processes, as well as medical procedures like surgery, cannot simply be halted but for the price of catastrophic consequences.

The complexities of medical work and the inherent ambiguities which attend medical practice foreclose any assumptions about reducing uncertainty and the need for improvisation completely. Ethnographies of medical work subsequently highlight not only a chance of improvisation, but an imperative for improvisation. This imperative for improvisation is augmented in high-tech settings by having to deal not only with technical malfunctions but also with organisational disruptions. It would indeed be naïve to assume that we necessarily find less improvisation in a modern high-tech hospital than in a nineteenth-century field lazaretto. Rather, we find different modes of improvising. Improvising under the conditions of scarce resources is different to improvising in affluent situations, but both share the same background of adjusting to unforeseen situations depending on the means at hand. Most of the anaesthetists I talked to during my ethnographic observations were actually not at ease with me referring to their work as improvisation. Some also talked of the need for improvisation in emergency situations, but they would not readily speak of their daily practice as improvisation. They feared that this might lead to misunderstanding their work as incompetent practice which is not my intention at all.

As a sensitising concept for studying medical practice, improvisation denotes this situated assembling of heterogeneous elements in an ad hoc fashion. Improvisation and related terms, such as tinkering or bricolage, have been used regularly to emphasise the emergent characteristics of organised work and repair practices (Knorr 1979; Orr 1996; Crossan 1998; Henke 2000; Garud and Karnøe 2003; Graham and Thrift 2007) even of social practices themselves (Bourdieu 1977 [1972]). Henke (2000: 66) explicitly stresses the situated nature of improvisation as a defining characteristic of repair. This understanding of improvisation essentially extends Blumer's notion of a sensitising concept in the way that it does not only guide the empirical gaze, it demands that improvisation as situated and distributed action must be studied in practice.

Improvisation and Skilful Practice

The second point is related to the first. The ad hoc and situational character of improvisation should not be conflated with inferiority or incompetence. To improvise does not mean to be unskilled, incapable, or without resources. On the contrary, competent situational adjustments require a high level of training, experience and skill (Pinch et al. 1996). In line with the sociological reasoning on jazz improvisation (Sudnow 1978; Becker 2000), improvisation as a sensitising concept emphasises that to improvise is not simply doing something without preparation, but an activity which requires a good deal of training, experience, and knowledge. It also entails that improvisation is not the creation of something entirely new, but a variation of existing standards. A variation, in which planning and performance collapse within deeply situated activities. The third vignette showed that improvisation in medicine also cannot be separated from questions of experience and authority.

Yet, in contrast to musical improvisation, medical improvisation is not an end in itself, not the opus operatum, but the modus operandi dealing with unexpected disruptions of the workflow. It is part of the mostly invisible articulation work in medicine (Strauss et al. 1985: 151), a necessary part of the negotiated order in hospitals.

Skilful and competent improvisation thus lies at the heart of a practice-oriented and pragmatic understanding of the creative aspects of human action (Ryle 1976; Joas 1996; Sawyer 2000). As a modus operandi of repair work, improvisation lead us to think of medical practice not as the application of pure scientific knowledge or technical procedures, but as a constant process of competent situated variation and verification, of making sure and staying on course (Schubert 2011). As a sensitising concept, improvisation then points to the interplay of routine and disruption, of breakdown and repair in medical practice. In this sense, it more often than not means to habitually create workarounds, to switch courses of action, to replace broken devices with ones that work.

Improvisation and Socio-Technical Ensembles

In addition to the situated and skilful aspects of improvising, improvisation as a sensitising concept brings human–technology interrelations to the fore (Schubert 2012). Jazz improvisation illustrates this point nicely. However, in opposition to jazz improvisation, repair improvisation extends past the skilful mastery of musical instruments, as it denotes the skilful dealing with more or less unruly technology. Improvisations with unruly instruments may be found in more experimental forms of music, but the instruments are nonetheless expected to create unexpected sounds.

Improvisation also extends past individual embodied conduct. It is a distributed social activity that is learnt through experience and interaction. Improvisation as a sensitising concept should therefore lead us to the question how improvisational and repair work circulate within medical practice. Doctors not only learn to handle the unforeseen disturbances in practice by embodying the necessary repair skills through experience and training. They also learn from more experienced doctors and to a great extent also from the nurses. Similar to the war stories of photocopy repair work that Orr (1996: 125) observed, medical repair knowledge is circulated inside and outside the operating room by telling stories (cf. Montgomery Hunter 1991; Atkinson 1995)

Conceiving improvisation in this way transgresses the boundaries of a given situation, it reaches past situational conduct and embodied skill. The technical arrangements of modern operating rooms, for instance, afford specific forms of improvising. The competence of doctors and nurses includes their past experiences. Ideologies of improvisation regulate which modes of repair are acceptable and legitimate. In this extensive understanding, improvisation not only links humans and non-humans. It also links the past and absent with the present, the routine with the deviant.

Using improvisation as a sensitising concept for studying medical practices in the way I have just framed it, opens up the complexity of repair work in a specific way. In short, it emphasises the situated, skilful, and distributed nature of repair and medical practice alike. As a sensitising concept, I have already refined it "in the light of stubborn empirical findings" (Blumer 1954: 8) within medicine. I have set it apart from jazz improvisation and popular notions of improvisation as makeshift botching. Improvisation, it must be concluded, is not uniform. Medical improvisation is different from musical improvisation (Becker 2000), organisational improvisation (Kamoche et al. 2003), improvisation in street-selling (Clark and Pinch 2014 [1995]), improvisation in computer modelling (Landström et al. 2013), or improvisation in teamwork (Middleton 1996), because the lives of patients are at stake and because it is situated in specific constellations of technologies, organisations, and bodies.

In sum, repair work and improvisation are abundantly found in medical practice. As I have tried to show, they can also be used as conceptual approaches for analysing the relations of breakdowns and repair as well as relations of humans and non-humans. Medical practice might be a curious case in this respect, because repair does not necessarily imply fixing a broken machine, but rather working around safety features, deactivating fully functional devices, or simply switching machines on and off.

One further curiosity needs to be added. Stressing the skills and situated activities of medical repair work leads to an action-bias in the conceptualisation of medical work and the questions is, if the contingencies of modern medical practice always have to be dealt with by taking action. In certain cases, the underlying interrelations between bodies and technologies are obscured through ambiguous causes and symptoms. Experienced anaesthetists thus do not have a purely interventionist attitude, but tend to be careful not to engage in hasty repairs. Sometimes, they wait and see how the situation develops in order to find out what it actually is that needs fixing—or as the quote at the beginning of the chapter phrases it: "you have to know a lot to do nothing".

Conclusion

This chapter aimed at empirically and conceptually disclosing the multiplicities of medical repair work by drawing on ethnographic observations, pragmatist philosophy, and interactionist sociology. The main themes revolved around the ordinary repair practices of doctors and nurses as they routinely and skilfully deal with technical dysfunctionalities and organisational disruptions. My treatment of the issues was connected through questions concerning their specificity and generalisation, the relation of routines and disruptions, and the relation of humans and artefacts. Despite these common themes, the paper does not fall neatly into place by concluding into one coherent picture. This would be presumptuous in light of the empirical diversity. At the same time, it does not remain attached to singular situations. Rather, I aim for an empirical and conceptual collage that mobilises different relevances (Kalthoff 2010: 363). In other words, I argue for a comparative perspective that does not try to settle the issue of repair in medicine, but for one that actively seeks to open up the issue in medicine and beyond.

On the one hand, I argued for a more general understanding of medical repair work as practical activity in terms of inquiry. On the other hand, I have emphasised the peculiarities of modern medical practice both from my own research as well as from the sociological literature on medical work. Repair, it was argued, is not only a fundamental feature of practical action per se, but also plays a prominent role as collective activity in the sociology of medical practice. This necessarily entails a strong notion of repair en passant—as interactive and ongoing accomplishment (Garfinkel 1967), in which repair is not excluded from the normal flow of time or happens in a knife-edge present, but unfolds in the situated transforming of an indeterminate into a determinate situation (Dewey 1938).

Likewise, using improvisation as a sensitising concept for observing and analysing medical repair work can help to connect repair studies of medical work with repair studies in other fields. The situated, skilful, and distributed nature of improvisation resonates well with a more general interest of repair studies in the situated maintenance and care of socio-material associations. The curiosity of medical repair work might not be so curious after all, but only one empirical example of the manifold ambiguities associated with technical work in contemporary societies.

Note

1. The complexities of breakdown and repair in organised medicine touch on many issues, responsibility being one that has not been made prominent yet. I will not go into more details of medical repair work and responsibility at this point—only as far as it concerns the concept of improvisation (cf. Becker et al. 1961 for an excellent discussion of responsibility and experience in medical training).

References

Atkinson, Paul A. 1995. *Medical Talk and Medical Work: The Liturgy of the Clinic*. London: Sage.

Becker, Howard S. 2000. The etiquette of improvisation. *Mind, Culture, and Activity* 7 (3): 171–176.

Becker, Howard S., Blanche Geer, Everett C. Hughes, and Anselm L. Strauss (eds.). 1961. *Boys in White. Student Culture in Medical School.* Chicago: University of Chicago Press.

Berg, Marc, and Annemarie Mol (eds.). 1998. *Differences in Medicine: Unraveling Practices, Techniques, and Bodies*. Durham: Duke University Press.

Blumer, Herbert. 1954. What's wrong with social theory? *American Sociological Review* 19 (1): 3–10.

Bosk, Charles L. 2003 [1979]. *Forgive and Remember: Managing Medical Failure.* Chicago: University of Chicago Press.

Bourdieu, Pierre. 1977 [1972]. *Outline of a Theory of Practice*. Cambridge: Cambridge University Press.

Bruni, Attila, Trevor J. Pinch and Cornelius Schubert. 2013. Technologically dense environments: What for? What next? *Tecnoscienza* 4 (2): 51–72.

Clark, Colin, and Trevor J. Pinch. 2014 [1995]. *The Hard Sell. The Art of Street-Wise Selling.* London: HarperCollins.

Collins, Harry M., Gerard H. de Vries, and Wiebe E. Bijker. 1997. Ways of going on. An analysis of skill applied to medical practice. *Science, Technology & Human Values* 22 (3): 267–285.

Crossan, Mary M. 1998. Improvisation in action. *Organization Science* 9 (5): 593–599.

Cubasch, W. 1884. *Die Improvisation der Behandlungsmittel im Kriege und bei Unglücksfällen. Vademecum für Ärzte und Sanitätspersonen.* Wien: Urban & Schwarzenberg.

Dant, Tim. 2010. The work of repair. Gesture, emotion and sensual knowledge. *Sociological Research Online* 15 (3).

Denis, Jérôme, and David Pontille. 2015. Material ordering and the care of things. *Science, Technology and Human Values* 40 (3): 338–367.

Dewey, John. 1916. *Essays in Experimental Logic*. Chicago: University of Chicago Press.

———. 1930 [1929]. *The Quest for Certainty. A Study of the Relation of Knowledge and Action.* London: Allen & Unwin.

———. 1938. *Logic: The Theory of Inquiry*. New York: Henry Holt.
Fox, Renée C. 1957. Training for uncertainty. In *The Student-Physician. Introductory Studies in the Sociology of Medical Education*, ed. Robert K. Merton, George Reader, and Patricia L. Kendall, 207–241. Cambridge: Harvard University Press.
Freidson, Eliot. 1970. *Profession of Medicine. A Study of the Sociology of Applied Knowledge*. New York: Harper & Row.
Garfinkel, Harold. 1967. *Studies in Ethnomethodology*. Englewood Cliffs: Prentice-Hall.
Garud, Raghu, and Peter Karnøe. 2003. Bricolage versus breakthrough: Distributed and embedded agency in technology entrepreneurship. *Research Policy* 32 (2): 277–300.
Gießmann, Sebastian, and Gabriele Schabacher. 2014. Umwege und Umnutzung oder: Was bewirkt ein "Workaround"? In *Umnutzung. Alte Sachen, neue Zwecke,* ed. Stephan Habscheid, Gero Hoch, Hildegard Schröteler-v. Brandt, Volker Stein, 13–26. Göttingen: V&R unipress.
Goodwin, Dawn. 2007. Upsetting the order of teamwork. Is 'the same way every time' a good aspiration? *Sociology* 41 (2): 259–275.
Graham, Stephen, and Nigel Thrift. 2007. Out of order: Understanding repair and maintenance. *Theory, Culture & Society* 24 (3): 1–25.
Harper, Douglas. 1987. *Working Knowledge: Skill and Community in a Small Shop*. Chicago: University of Chicago Press.
Heath, Christian, Paul Luff, and Marcus Sanchez Svensson. 2003. Technology and medical practice. *Sociology of Health & Illness* 25 (3): 75–96.
Henke, Christopher R. 2000. The mechanics of workplace order. Toward a sociology of repair. *Berkeley Journal of Sociology* 44: 55–81.
Hughes, Everett C. 1951. Mistakes at work. *Canadian Journal of Economics and Political Science* 17: 320–327.
Jackson, Steven J. 2014. Rethinking repair. In *Media Technologies. Essays on Communication, Materiality, and Society*, ed. Tarleton Gillespie, Pablo J. Boczkowski, and Kirsten A. Foot, 221–239, Cambridge: MIT Press.
Joas, Hans. 1996. *The Creativity of Action*. Cambridge: Polity Press.
Kalthoff, Herbert. 2010. Beobachtung und Komplexität. Überlegungen zum Problem der Triangulation. *Sozialer Sinn* 11 (2): 353–365.
Kamoche, Ken, Miguel Pina e Cunha, and João Vieira da Cunha. 2003. Towards a theory of organizational improvisation: Looking beyond the jazz metaphor. *Journal of Management Studies* 40 (8): 2023–2051.
Knorr, Karin. 1979. Tinkering toward success: Prelude to a theory of scientific practice. *Theory and Society* 8: 347–376.

Landström, Catharina, Sarah J. Whatmore, and Stuart N. Lane. 2013. Learning through computer model improvisations. *Science, Technology and Human Values* 38 (5): 678–700.

Middleton, David. 1996. Talking work: Argument, common knowledge, and improvisation in teamwork. In *Cognition and Communication at Work*, ed. Yrjö Engeström, and David Middleton, 233–256. Cambridge: Cambridge University Press.

Mol, Annemarie, Ingunn Moser, and Jeannette Pols (eds.). 2010. *Care in Practice. On Tinkering in Clinics, Homes and Farms*. Bielefeld: Transcript.

Montgomery Hunter, Kathryn. 1991. *Doctors' Stories. The Narrative Structure of Medical Knowledge*. Princeton: Princeton University Press.

Mort, Maggie, Dawn Goodwin, Andrew F. Smith, and Catherine Pope. 2005. Safe asleep? Human-machine relations in medical practice. *Social Science and Medicine* 61 (9): 2027–2037.

Mort, Maggie, Carl R. May, and Tracy Williams. 2003. Remote doctors and absent patients. Acting at a distance in telemedicine? *Science, Technology & Human Values* 28 (2): 274–295.

Orr, Julian E. 1996. *Talking About Machines: An Ethnography of a Modern Job*. Ithaca: ILR Press.

Parsons, Talcott. 1951. *The Social System*. New York: Free Press.

Pinch, Trevor J., Harry M. Collins, and Larry Carbone. 1996. Inside knowledge: Second order measures of skill. *The Sociological Review* 44 (2): 163–186.

Popitz, Heinrich, Hans Paul Bahrdt, Ernst August Jüres, and Hanno Kesting. 1957. *Technik und Industriearbeit. Soziologische Untersuchungen in der Hüttenindustrie*. Tübingen: Mohr Siebeck.

Ryle, Gilbert. 1976. Improvisation. *Mind* 85 (337): 69–83.

Sawyer, R. Keith. 2000. Improvisation and the creative process: Dewey, Collingwood, and the aesthetics of spontaneity. *The Journal of Aesthetics and Art Criticism* 58 (2): 149–161.

Schubert, Cornelius. 2006. *Die Praxis der Apparatemedizin. Ärzte und Technik im Operationssaal*. Frankfurt am Main: Campus.

———. 2007. Risk and safety in the operating theatre: An ethnographic study of socio-technical practices. In *Biomedicine as Culture: Instrumental Practices, Technoscientific knowledge, and New Modes of Life*, ed. Regula V. Burri, and Joseph Dumit, 123–138, London: Routledge.

———. 2011. Making sure. A comparative micro-analysis of diagnostic instruments in medical practice. *Social Science & Medicine* 73 (6): 851–857.

———. 2012. Distributed sleeping and breathing: On the agency of means in medical work. In *Agency Without Actors? New Approaches to Collective Action*, ed. Jan-Hendrik Passoth, Birgit Peuker, and Michael Schillmeier, 113–129. London: Routledge.

Strauss, Anselm L. 1978. *Negotiations*. San Francisco: Jossey-Bass.

Strauss, Anselm L., Shizuko Fagerhaugh, Barbara Suczek, and Carolyn Wiener. 1985. *Social Organization of Medical Work*. Chicago: University of Chicago Press.

Strauss, Anselm L., Leonard Schatzman, Danuta Ehrlich, Rue Bucher, and Melvin Sabshin. 1963. The hospital and its negotiated order. In *The Hospital in Modern Society*, ed. Freidson, 147–169. New York: Free Press.

Suchman, Lucy A. 1987. *Plans and Situated Actions. The Problem of Human-Machine Communication*. Cambridge: Cambridge University Press.

———. 2007. *Human-Machine Reconfigurations. Plans and Situated Actions*. Cambridge: Cambridge University Press.

Sudnow, David. 1978. *Ways of the Hand: The Organization of Improvised Conduct*. Cambridge: Harvard University Press.

Timmermans, Stefan, and Marc Berg. 1997. Standardization in action: Achieving local universality through medical protocols. *Social Studies of Science* 27: 273–305.

———. 2003. The practice of medical technology. *Sociology of Health and Illness*, 25 (3): 97–114.

Wright, Peter, and Andrew Treacher (eds.). 1982. *The Problem of Medical Knowledge*. Edinburgh: Edinburgh University Press.

3

'A Good Enough Fix': Repair and Maintenance in Librarians' Digitization Practice

Moritz F. Fürst

Introduction

Standing on a ladder in the State Hall of the Austrian National Library in Vienna, a man dressed in a lab coat, gloves and a face mask, removes a book from the gigantic shelf with extreme care. He hands it down to a colleague, who gently places it onto a small cart, already laden with other volumes. Their actions suggest they are handling an unstable, contagious substance rather than mundane, everyday objects that are subjected to almost innate routines of usage (the ones that you, the reader, are employing right now). Yet, while this cautiously handled item may be two hundred or more years old, it could nevertheless be a book that a visitor might have taken to one of the desks in the adjacent reading room just a few weeks before, just like almost all the other volumes within the library's inventory. This part of the building is a museum, a carefully protected site of cultural heritage and national identity. But it

M. F. Fürst (✉)
University of Lausanne, Lausanne, Switzerland
e-mail: moritz.fuerst@unil.ch

I. Strebel et al. (eds.), *Repair Work Ethnographies*,
https://doi.org/10.1007/978-981-13-2110-8_3

is still a *library* after all, actively encouraging its visitors to dive into the endless depths of its gathered knowledge.

Shadowing librarians in their daily work routines exposes such seemingly paradoxical qualities of their objects of care on a regular basis. Durability and stability, properties commonly attributed to objects within studies of material culture (see Ingold 2007), apparently do not suffice to characterize *these* material objects. At the same time, nothing seems to be more "immutable" (Latour 1986) than the grandiose inventory the librarians refer to in talk, wherein every work has its particular place in a dense network of stories forming this particular 'historical collection.' How can we account for the socio-material arrangement at play here? What *are* librarians' objects?

To study such relations, it is perhaps less useful to start, as is the case within the dominant paradigm in library and information studies (LIS), with the question 'what is information'/'what is an information artifact' abstracted from the situated socio-material orderings they are embedded in (Cornelius 2004; Watson and Carlin 2012). Rather, this chapter proposes an alternative approach to the study of information artifacts by investigating librarians' work practices. Among the various tasks involved in providing the logistics for intellectual work, the librarians' profession entails taking care of their repository, not only in terms of continuing a bibliographic arrangement, but also in physical care of the material 'things' it contains. Like buildings, traffic systems or the power grid, the infrastructural underpinnings of knowledge and cultural heritage are held together by a set of activities "by which the constant decay of the world is held off" (Graham and Thrift 2007: 1), namely the practices of maintenance and repair.

The central aspiration of all archival work, the "catechontic" (Ernst 2002: 127) effort to introduce and sustain order, and suspend chaos among information artifacts, finds its mundane and very lively parallel in the struggle against the material disintegration of fragile objects—of books that are moved, flipped, yanked, dropped, torn, and thrown. The paramount and distinguishing feature of the library, however, lies not simply within the collection and preservation of such objects, but that they are made available for access (Cubitt 2006). The librarian is an archivist and conservator; yet, her main concern is to enable others to *use* their inventory: to identify the relevant, find the unexpected, and connect the item at hand with other materials. The techniques directed toward such a task (cataloging, classifying, sorting, etc.) perpetuate an arrangement beyond a mere conglomeration of information artifacts. Within the library, the whole is indeed more than the sum of its parts, and this 'whole' is continuously in the making through maintenance activities.

In recent years, digitization has been a central (if not the central, see Coyle 2006b) aspect of such activities. The proliferation of digital technologies in the everyday life of contemporary societies has confronted librarians with many challenges. The métier increasingly finds itself entangled within the transformative dynamics of contemporary knowledge regimes and their socio-material reconfigurations. In a time of 'digitization,' that is to say, the sense of an ongoing process of social, political, economic, and cultural life increasingly adapting to a "digital paradigm" (Ceruzzi 2012: x) of dealing with information, the documentary heritage artifact (or 'document' in the parlance of the profession) becomes problematic.

What exactly constitutes such an artifact is thus not solely a possible question for social theory, but a very *practical* question of expanding importance, in light of the bulk increase of media introduced by technological and social developments over the past two hundred years (Buckland 1997; Nichols 2006; Westin 2013). Whatever one might understand by libraries, what they are is closely tied to what materials they hold. The question, 'what are our objects of concern?' identifies and demarcates an object range for institutions which have long gone beyond 'books' as the only materials within their holdings. Which kinds of information artifacts should be collected, how they should be described, organized, made accessible, and protected for future use, are all pressing concerns for librarianship in very practical terms. With vast amounts of information produced every day, digitization in this sense amplifies the lack of consensus on what exactly makes up an 'artifact' to collect, house, and preserve within a particular institutional arrangement.

Furthermore, the adoption of digital technologies has consequences for the materials already housed within libraries. While 'surrogates' of information artifacts have, at least since the introduction of optical and magnetic reproduction technologies, been an integral part of maintaining the library's objective, the conversion of objects to digital data has taken the task to levels of an "industrial" (Coyle 2006a) scale.[1]

Investigating maintenance and repair practices illustrates how such issues are taken up and dealt with in the daily work of librarianship. It reveals how information artifacts are reified as historical documents, and how 'doing documentary heritage' is enacted in multiple ways of everyday talk and practice. This chapter seeks to elucidate repair and maintenance work through, in, and as digitization.

I will draw on an ethnography of librarians' work conducted at the Austrian National Library, currently in the process of digitizing its complete historical holdings from the sixteenth to the nineteenth century. Exemplifying a few of the manifold ways in which librarians relate to their objects, this text will propose an empirical account of the various associations through which activities, procedures, tools, buildings, and bodies are mobilized in a digitization endeavor. Drawing on conceptions developed in actor-network theory and post-ANT studies to probe 'ontological' matters in empirical terms, I will address the multiple 'enactments' of objects in the different sites of a library currently conducting a mass digitization project.

Investigating librarians' activities through the lens of repair and maintenance work challenges the divide between production and use of (digital) information found in the predominant discourse of innovation, design and transforming social potential in media and information research (Jackson 2014). Furthermore, it reconsiders socio-material relations beyond a dualistic 'analog-digital' conception of objects and their status in the making of the social.

Assessing Damage

Like built heritage monuments or art works, written records of deemed cultural importance are subjected to conservation practices. Ensuring the continued usable state of their materials is an integral part of librarians' care taking. The work of the conservators is defined by the local setting, and by the specific corpus of objects they work with. The conservation team talks about this corpus in different terms to members in other areas of the library and people working on the mass digitization project.[2]

In my interactions with them and in conversations I witness, they refer to the book *solely* in material terms, with no mentions of valuation based on other criteria. An 'old' book, such as the ones being scanned in this digitization project, is not described as 'old.' Neither is it discussed along criteria of 'historical value as a cultural heritage object,' or by joking about its seemingly absurd title. It is rather talked about as having distinct material qualities typical to a certain manufacturing style, or a fragile, vulnerable thing that needs to be prevented from disintegrating. Talk about sub-collections involved in the digitization project is organized around the (in this case worse) 'conservation state' compared to other parts of the overall holdings; or the average physical 'thickness' of individual volumes compared to other subcollections.

The 'single' object is predominantly only of secondary concern. What matters is the corpus, and the relation between the individual objects that constitute it. This arrangement is volatile. It is a disintegrating state of affairs which needs to be stabilized. In general, monitoring the damage of such a large quantity of books is elaborated guess-work, against a big unknown body of objects constantly moving toward *further* decay. An 'assessment of the conservation state' in day-to-day operation is conducted on a per-book basis (e.g., when a reader requests it), accompanied by scheduled routine assessments of subparts of the collection; often not on an individual per-book basis, but by taking samples of books deemed to be in a similar 'physical shape.' Basic documentation (the numbers of conducted repair tasks in categories such as 'loose pages,' 'fissure on cover') serves as a further indication for the state of things. Individual documentation of the conservation state of a particular object (e.g., in the internal catalog) is not common and disregarded as impractical due to the sheer size of the collection. The conservation expert can thus 'get a picture' of the overall state, without necessarily having 'hard facts' on her *whole* inventory at any given moment.

Only at very few points in time, is a so-called 'full revision' or '*Schadensbildaufnahme*' is practical (described to me as a 'twice in a lifetime' event by members of this team)—a time-consuming process wherein every single object is individually assessed. The *Austrian Books Online* project is such an event for the conservation team. As such, this digitization is not just about creating digital representations; for the conservators, it is to a greater degree simply an event to learn about their corpus and its characteristics—the material 'shape' it is in.

What is the 'conservation state' of a book, or a collection of books? Just when is an old book considered to be 'in bad shape'? What are the properties to evaluate this condition? How are they 'assessed' in practice? For answers to questions such as these, it is useful to consider the kind of repair tasks that are routinely conducted. An instantly identifiable feature of the 'conservation state' from members' talk and how they organize everyday work, lies in the fact that it seems to be organized around two principles.: The fact *that* there is recognizable 'damage' to the object, which is intelligibly different (even for nonconservators) from abrasion or wear; and that this damage can be *fixed* to return the object to a desired state of functionality. The condition of an object is described by referring to whether or not it has a *Schaden* ('defect,' 'damage') that falls into one of the categories of repair tasks, which members use in talking to each other and documenting their practices.

When visiting the workshop of the conservation team, inquiries into the repair tasks undertaken would result in more or less elaborate descriptions of 'fixing defects' of the following nature:

- loose jacket, damaged jacket
- fissure on/loose 'bonnet'
- fissure on/loose 'foot'
- broken spine
- loose pages, fissured pages
- loose *Titelschild* (title label)
- mold on the outer pages
- etc.

Members of the conservation team predominantly talk about *functional* aspects of their book objects, and only sparingly relate to *aesthetic/Gestalt* features (often only in conjunction with a fix to a 'functional' problem). While both of these categories exist in members' talk, the priority in dealing with the former kinds of problems becomes instantly clear. Members would say, "It's not about eliminating every bad spot, we just have to ensure that it survives the transport and the scanning and that nothing gets lost during the process." Such concerns resemble repair and maintenance practices described as directed toward restoring order (Henke 1999; Gregson et al. 2009). More specifically, they are an "attempt to arrest the traces" (Gregson et al. 2009: 248) of usage. The main objective, however, is not the elimination of traces of usage, but the protection of further decay. The 'integrity' of a work is ensured when it can be 'fixed' to a degree that no *Materialverlust* ('loss of material') occurs during its use or while it is stored; the goal is to secure the book so that it will not *disintegrate further.*

The process of assessing damage that counts as damage is considered to become 'apparent' during a 'reading occurrence' such as the one cataloging-personnel carry out when they take a book from a cart and open it to perform their work. Once they take a volume into their hands, open it, and quickly page through it, 'functional' damage can be recognized through the use-situation itself. Any dysfunction occurring during the reading occurrence indicates that there is indeed a problem. While their "professional vision" (Goodwin 1994) allows the conservators to assess a problem to different and greater levels of detail, they attribute 'lay persons' with the ability to identify that a problem *exists*, or conversely that no problems exists for a given object.

Furthermore, the use-situation is described by members as performing an assessment of the anticipated demands on the material during their subsequent journey in the digitization project. Conservators explain to me that modern book-scanning technology allows for a procedure which puts a physical strain on the book comparable to an ordinary 'reading occurrence.' The book is opened to a maximum angle of 120 degrees and an operator flips the pages while an overhead camera takes pictures. Thus, from the point of view of the conservator, the negligible differences in how the material features are put under pressure result in these situations being of such a similar kind, that one can be used to evaluate the other; an object which can be *read* without problems can also therefore be *scanned.*

When the conservators get a returned batch from the scanning facility, they move the shipping carts back up to the manipulation room and perform a 'visual assessment' of their books. A conservator glances at the picture of the cart taken before the books were shipped to the scanning facility and compares it to what she has in front of her eyes. Her eyes move between photograph and cart as she checks if the order on the cart is right, if there are loose parts somewhere, if something is 'off.' After she is finished, she takes orange slips and places them inside the books that need care and repair work. This technique is intended to give the conservators a quick idea of possible 'structural problems' with a batch. Should certain problems arise repeatedly with books in the same batch, this information is not only useful to estimate the amount of necessary repair work, but also as an indicator that there is something wrong with the workflows at the scanning facility. Since the library and Google conduct the different tasks of the digitization process independently, the conservation team has no insight into the exact procedures and is not involved in 'minor' workflow changes at the scanning facility. However, such changes do have an effect on their work, since they have to ensure they can meet the schedule for preparing future batches. If they have to spend too much time on post-scanning repair, they will not meet this schedule.

On these grounds, digitization has a profound impact on how caretaking work is conducted in a rare instance of 'full revision,' both in terms of being its cause, as well as prompting routines that shape maintenance practice. This close interplay becomes even more apparent if one looks at the actual repair activities.

Temporary Restoration

The repair and conservation work that is carried out on library books is usually not aimed at achieving any form of pristine condition in aesthetic, nor in functional terms. Members talk about 'securing' and 'temporarily restoring,' not about 'repairing' in a more inclusive sense. The conservators would also point out how the 'fixes' they have applied to a book are often only temporary, 'good enough' solutions preventing further damage to the original materials, and are even expected to break at times. A 'complete re-assembly' is seldom desirable, even if it is considered to be a functionally or aesthetically better solution, and is avoided where possible.

For one, because as conservators said, it would "lower the status, the value of the object, its authenticity," by which they only vaguely mean its 'historicalness' is not to be tampered with, but predominantly its integrity as a material object crafted using certain techniques of book-making. The notion of authenticity performed within this context is thus a locally situated negotiation in and out of repair practice (Jones and Yarrow 2013).

Furthermore, time constraints do not allow for extensive restoration tasks on a large number of books, a factor further exacerbated by the tight schedules of the ongoing mass digitization project. The restoration work during the project is therefore conducted mainly in the general area of preparation for shipment to the scanning facility. It predominantly consists of what are considered to be 'routine' practices.

An exception is work intensive subcollections, prepared ahead of schedule down in the conservators' workshop when the normal daily workload allows for it. Furthermore, certain types of damage require the use of special instruments and environments (e.g., for removing mold). The 'routine' fixes and small repairs can usually be conducted with a small number of tools on a simple workbench in the preparation area. The conservators working here are mainly on their own, independent from what is going on down in the workshop, and fully devoted to working on the *Austrian Books Online* project. To perform restoration work, a conservator will take a book from a cart marked with a blue slip of paper, signaling that there is 'something wrong' with it, and take it to their desk. While they *assume* that there is a certain problem with the book, the repair task begins with a diagnosis of the problem, and a decision on what action to take. In relating to the book at hand in need of a quick fix, the irredeemably localized and contextually contingent nature of 'diagnosing damage' and 'deciding what to do' in book repair work becomes apparent. Demonstrating their work to me, conservators would make use of categorizations of an object, using the collection-as-a-whole as a 'resource' (see Zimmerman and Pollner 1970; Garfinkel and Sacks 1986).

Pointing toward the lower point of the spine of a book, telling me it is "one of Prince Eugene's books, they have a particular jacket and always break here," the conservator produces an ad hoc categorization of this particular book as being part of a corpus of works that 'always' need repairing in the same place. He does so by relating it to a subpart of *this particular collection the library holds.* That this is 'one of Prince Eugene's books' serves as a contextual feature of deciding what to do. These categorizations are furthermore used to define what a 'routine' repair task is; since everything beyond routine is to be excluded and 'pushed in red' (marked with a red slip of paper signaling exclusion from the digitization process) to be worked on at a later time. Again, the ultimate goal is to prevent further damage; postponing a repair can be an alignment with that goal, on the cost of declaring it 'defunct' for usage (by readers or a scanner). This tension between preservation as a heritage object (the legal obligation of this national library) and functional capability for usage is the result of everyday librarians' practice; neither a discursive, social construction, nor an immanent feature of 'objects.'

The Book Multiple

The collection the conservators attune to is not a collection of beautiful museum objects with a rich, intertwined history. They relate to the same books, but their objects are messy assemblages of leather, paste, paper, thread, dust, mold, glycerin, integrity, fragility, craftsmanship, and decomposition. They are conscious of their objects' 'histories,' but that only matters in the pragmatics of getting objects functional again, or preventing them from being further damaged. The stories they tell about the 'historical collection' are ones about "poor jacket quality of 19th century books."

These stories differ substantially from the ones told by librarians mainly concerned with cataloging and metadata-management work on the collection. Their object of care is the book embedded in circumstances of whole empires, peoples, geographies, traditions, cultures, laws, relations, conversations, scholarship, provenience, and accumulation. Their goal lies in improving its descriptions, and supporting others who do so. Paying attention to how their books were described in the past, how they ended up in the holdings of this library, and how these holdings form a corpus, characterizes their work. As such, digitization of their inventory is framed in *adding* yet another description, another representation of the object to the ones that already exist within this library.

Members' talk emphasizes the collection as a cohesive, orderly whole over a mere large accumulation of old, rare or unique works. Forming the collections' consistency is achieved in talk by differentiating the corpus into subdivisions and relating them to each other, introducing historical continuities of the libraries' purpose and activities and providing rationale for their 'place' within the whole collection. There are the books assembled by individuals (such as Prince Eugene of Savoy); there are different corpora of works not written in German, rooted within the *Pflichtablieferung* (a 'legal deposit' clause to provide the library with copies of every work printed) which was established in the sixteenth century during the Habsburg monarchy; there is also the fact that the library was oriented toward 'universal' criteria until 1920 as opposed to the 'Austrian' focus continued/adopted afterwards.

Yet, it is not simply talk that keeps the objects in question distinct. The setting resembles themes congruent with recent trajectories in science and technology studies (STS) to extend the displacement of epistemology's grand themes "to the realm of the ontological" (Woolgar and Lezaun 2013: 322). A core leitmotif of 'Post-ANT' (Law and Hassard 1999; Gad and Bruun Jensen 2009) propositions to subject ontological topics to empirical treatment is to call attention to the multiple ways of the world no longer in terms of representations, but to trace and describe different practical enactments of 'reality' in different situated instances.[3]

However, in studying librarians' work, such 'multiplicity' continuously emerges within the field prior to any analytical postulate. The socio-material matters identified by the ethnographer during the course of investigating 'librarians' work' are in part inherent in participants' own reasoning toward their work practices: doing ethnography reveals salient instances wherein the analyst's "constructs of the second degree" (Schütz 1971 [1953]: 7) find their parallel in librarians' vernacular theory of what it is they do. The 'multiple associations' between the 'social' and the 'material'/'objective' are, so to speak, often made for the ethnographer by those he studied, during and as part of their everyday affairs. As such, different 'versions' (as referred to in the parlance of a general ontology of multiplicity) of objects, the required work to make them coherent in different parts of the library and the digitization project, are, independent of the analyst's 'intervention,' an integral part of the practices encountered.

Consider the following excerpt, touching upon discovery as a stabilizing by-product of mass digitization. An 'artifact' always derives its stability through the constant maintenance of the catalog. A digital version of the book thus does not 'change' the heritage object. It adds a new dimension of how it is described, it makes it even more stable, even more immutable. Digitization enables a new way of exploring what constitutes the library's holdings. Additionally, the project of digitization itself is an opportunity to explore the corpus. Both of these aspects are raised in the following quote, where a librarian talks about what she calls her "found pieces" (*Fundstücke*). By this, she means books that her team has found to be of particular value *as a result of* being resurfaced by the digitization project. This quote is taken from a casual conversation with the ethnographer about the holdings and the large amount of objects these holdings consist of.

> So here I have like a whole box. I call that found piece, because we partly don't know that we so- That often are like small, thin booklets. That it is actually unique. That nobody … at least not knowingly, yes? We don't know it, nobody except from us has this work. ABO serves. It's not just about opening with the scans, but also it serves protection because we now know we have to extra protect it. We suddenly discover what we have here. And we can protect it.

The expression of 'protection' is interesting because it diverges from the 'major consensus narrative' expressed in official documents, which is also a narrative presented at large within the digital department of this library, namely that scanned images serve a 'protective' function in case of a catastrophic incident destroying the 'original' material.

The librarian does not take up the narrative in this instance, but instead offers a different conception of the function of a scan in the role of 'protection.' The electronic book serves as a means of discovering the 'uniqueness' of the *physical book.* As she explicitly states, "it's not just about opening" the collection to readers from the outside, but also for the caretakers themselves. The scans are made relevant for her own work; yet, she decisively emphasizes protection in the sense of protecting the object as a unique object; not in terms of protecting the object in other ways and according to other valuations. However, she also acknowledges the issue of 'opening' the holdings to the outside, and it being an important, or *the* important part of the library's endeavor of provision ("not *just,*" my emphasis).

The mention of "small, thin booklets" instructs her conversation partner to recognize the significant function of the digitized book in enabling the discovery of even the most seemingly 'unremarkable' objects that she would 'miss' in her own modality of practicing documentary heritage. In doing so, the different 'functions' are offered to be understood in terms of a *unified* enterprise; while at the same time establishing the difference of the things in question. Similar motifs of digital books' function of 'discoverability' are expressed within this team by telling stories about a researcher from the other side of the world expressing interest in the physical book after having discovered its unique properties via electronic means, or by hypothesizing about how the physical book in the collection will become 'more visible' due to its electronic representation being online for the world to see.

This quote also exemplifies the reflexive attuning to identity and difference of the objects in question. The ethnographer encounters several instances of talk where the differences of objects of librarians' care (books, scans, text, and data) are explicitly made a topic of conversation. While these different 'things' are subjected to different practical concerns within the library, they are *all* nevertheless routinely raised in instances of talk not part of the situational context of a particular practical concern at hand: people engaged in book cataloging talk about a scanned image; as people engaged in overseeing quality control of scanned images talk about cataloging books. The way in which they do so is interesting, because it is recurrently taken as an opportunity for reflecting upon an 'own thing,' 'the other thing,' and their relation to each other.

The most conspicuous occurrence of such reflection shows itself where the respective 'function' of the things in question is compared to each other (such as in the excerpt above). In doing so, members express their understanding of differentiation between the multiple ways in which a thing can be used and how it relates to the other things. For example, the motif of 'discoverability' is thematized when, another member of this team contests the "commonly-held view that a serendipitous find is only possible by browsing bookshelves" in highlighting what he thinks the electronic book is good for. Interestingly, such a "commonly-held view" can be found on the other side of the building, within the digital department, where the benefits of "what is possible only when we have full text [of the books able to be processed by software]," namely a "more detailed way to find what you want" is emphasized over the—in this instance, "mere"—serendipity as a result of only being able to vaguely assess the content of a book by its bibliographic metadata and subject classification. Identity and difference of objects, and their relation to practice, are a concern for librarians—an *explicit* concern. The inherent reflexivity toward such questions is especially relevant to one of the key elements of a large corpus of artifacts: maintaining order.

Maintaining Descriptions

Producing consistent descriptions is an important device for establishing and maintaining internal order within libraries' holdings. It is not only a precondition for usage of materials, but also plays an important part in maintaining heritage. Descriptions form the material, explicit expression of how individual works relate to each other, at present and in times past. Descriptions—which usually take the form of a written account—can *represent* an item. This method of meta-information has a particularly useful benefit. Its material record can be kept separate from the item itself. Thus, a *catalog* of such meta-information can be compiled. This representational codex, a layer of abstraction (or a 'gloss' if you will), is able to give some indication of all the items that make up a particular corpus, *absent of the object at hand.*

In this library, any book will have a multitude of such representations:

- It is described in an old, handwritten catalog (which, of course, is in itself an *object* of the library now). This description almost appears 'unordered' when one is used to the metadata customs of our time. It is, however, often quite elaborate, for example, featuring the provenience of the book, how the library came to acquire it, and a description of its material features.
- It is described in an abridged, typewritten transcription of this handwritten catalog on index cards, made in the 1960s. This 'mechanized' catalog is standardized in the meta-information it features. At the same time, it is not as comprehensive as the individual book concerned, since additional descriptions beyond formal bibliographical data (author, title, imprint, and signature) are not included.
- It is described in a second catalog based on typewritten index cards, organized around content-based meta-information (keywords). This representation describes its relation to other books or items that store information.
- It is described in a relational database that is based on the older catalogs; primarily on a digitization (scanning and OCR) of the typewritten *Zettel*-catalogs. The basis for the database currently in use was a scanning project in the 1990s; additional descriptions get added continuously, also taken from the old handwritten catalog.
- Finally, it is described in a catalog for users of the library which can be accessed online. The library is currently integrating all its different catalog databases so users can retrieve results for all of them via a unified search mask ("Quicksearch").

These different layers of descriptions enable the library to function in a fundamental way, and they give a historical account which is an integral part of the cultural heritage the holdings represent. However, achieving consistency is bound to continuous efforts.

To ensure the order of things within a library, the book and its representations have to match in terms of the state of affairs. There are instances when this relation can become a problematic conflation point of different enactments of the objects of care. This problem emerges when representations facilitating order are put into question, which is continuously the case in the mass digitization project. The first important step when books are pulled from the shelves for digitization is to equip them with a barcode linking them to their description within the catalog database. However, this process is not only ordered by representations, but also *shaping* them. After a book is pulled from the shelf and loaded onto a cart, it is checked against the catalog, and its *description* is updated. On the one hand, this is a step in a broader continuous effort of the library to establish the electronic database as the canonical, consolidated, comprehensive repository of descriptions. At this stage, the most important representation of the book can be supplemented, corrected, and further standardized and systematized. Inconsistencies, which have occurred as a result of the multitude of descriptions that exist of a book within the library, can be removed.

On the other hand, having 'correct' metadata is considered to be a 'mission-critical' requirement, in terms of process. The personnel have approximately ten minutes per book to keep up with the schedule of a certain batch prepared for transport; problem cases that would take a longer amount of time are immediately postponed. In the processes that follow, these works are the exclusive responsibility of another team. The strong emphasis on 'getting it right' when it comes to the unambiguous structure of metadata-sets observable at this time is a direct consequence of the digitization process. While a librarian is used to dealing with historically grown inconsistencies and multiple trajectories of how catalog entries describe books, computer programmers at the other end of the chain have confronted them with a different sense of 'metadata,' in how they construct the relation between the digital object and its respective description. The metadata-adjustments meet a demand set by the 'end product': the electronic book and the logic of its processing.

For example, a work consisting of multiple volumes might only have one entry within the catalog[4]:

Title: Schillers Werke.
Person/Institution: Schiller, Friedrich
Place of publication/Publisher: Stuttgart/Cotta

Year of publication: 1867
Language: German
Description: Footnote: Vol 1–10.

This is perfectly suitable for a reader. She is interested in *Schillers Werke*; and the 'description' that this is a multi-volume work suffices. She can tell her librarian she needs 'Vol 1,' or 'the first one,' or perhaps even 'the one with the poems.' On the shelf, the individual material objects that coherently and intelligibly make up *Schillers Werke* sit right next to each other, so the librarian will not have a problem selecting the desired volume. However, these pieces of information have to be provided in a *structured* way in order to be machine-readable. When later on in the process a physical book is scanned, the images created have to be linked to the corresponding catalog entry. Therefore, separate (sub-) entries have to be created for each particular volume of *Schillers Werke*, so a data object can be unambiguously linked to a set of metadata. Data objects cannot 'sit next to each other' in the same way physical objects can. In order for a machine to 'understand' the relations between them, these relations have to be *made explicit*. The creation of the electronic book therefore affects the description both of itself, as well as of its physical counterpart. The whole logistical chain of the scanning process and the logics of data-processing are embedded into the canonical description of the book in the library in this way.

This discrepancy in the understanding of 'descriptions' is a source of tension, a point where conflicting, coinciding enactments have to be coordinated—and the members of this library are very aware of that. Beyond situated instances of referring to the objects of 'documentary heritage' when they relate to multiplicity, they also explore the nature of their profession in terms of cultural differences within the organization in epistemic dimensions. A striking example of assessing not only distinctive expert knowledge, but also assembling it around modes of conduct and technology consists of a member of the digital library pointing to her computer screen saying, "there is a different *Denkkultur* [between computer scientists and librarians with "an affinity for technology" and "classical" librarians oriented towards Humanities-scholarship]." This vernacular, and, if one may say so, indeed very apt glossing over of the concept of "epistemic cultures" (Knorr-Cetina 1999) to denote the "machineries" through which programmers write software formed by a particular understanding of how information artifacts can be meaningfully represented in contrast to how the librarian of book-culture understands the issues, was followed within the conversation by the identification of areas which would require "mutual learning from each other." Through instances such as this, the nature of their profession is questioned by members and explanations that thematize 'socio-technical reconfigurations' (new technologies, new skills, different kinds of expertise, shifts in requirements formulated by users, etc.) are offered to account for such questions to be posed. The 'multiplicity' and sometimes conflicting nature of the objects of care, how different practices are related to them, and that they need to be coordinated somehow, are deliberations librarians are confronted with in their day-to-day work, and which they reflexively take up in organizing that work.

Through maintenance of the catalog, books retain their 'historicity' and their particular constellation is stabilized and further reinforced. At the same time, the practice of creating 'better' (more extensive, standardized, machine-readable, etc.) representations of the inventory ensures the continued key operation of the library: providing access to its materials. Digitization plays into both dimensions. What librarians, according to a classic joke, do not care about—the content of their books—becomes 'mobile' with the introduction of digital surrogates of substantial parts of their inventory. The establishment of new orderings of knowledge emerging around the book as "data object" (Rieder 2013) is already having consequential repercussions within the world of libraries and creates numerous new challenges (see Bishop et al. 2003; Evens and Hauttekeete 2011). At the same time, digitization reifies, and even redefines the 'original' (see Latour and Lowe 2011).

Conclusion

Librarians' work is directed toward collection, housing, order, provision, and preservation of documentary heritage artifacts ('documents'). Such artifacts are, like any other artifacts, social. Documents are an integral part of making and maintaining social order; yet, they themselves are part of that order and therefore need to be attended to, and cared for (Levy 2003). This chapter has explicated two sets of such care and maintenance practices: physical maintenance of books, and organization and maintenance of descriptions.

I have described the librarians' object and corpus in terms of multiple enactments (Mol 1999, 2002). In studying librarians' work at the different sites of the Austrian National Library, I have shown the performance of socio-material order being distinctive to this arrangement engages different kinds of objects.

A 'historical' one, embedded in the circumstances of collection work of this library. It is set apart from the book-to-be-read by its character of aesthetic value and significance on its own—it is not sourced for the textual 'information' it contains. Its readers are not so much interested in *the* Confessiones of Augustinus as they are interested in *these* Confessiones of Augustinus, this particular object at hand. The object itself is the story, not its content. Its character of 'immutability' derives its stability from a dense network of tales of its 'mobility' over centuries, its provenience; descriptions which are collected and being added by the librarians themselves, and in their support to others who do so too. These stories form a corpus. The corpus of the historical object is of an emergent solidity, as its descriptions are becoming increasingly dense.

Conversely, the corpus relating to the material object is characterized by steady movement toward decay. The material object is fragile and vulnerable, enacted in practices of caretaking through constant routines of repair and restoration work. These characteristics stand in contrast to the predominant discussions of materiality within STS stressing stability and durability (see Denis and Pontille 2015).

The coexistence of these objects eventuates in a corresponding 'digitization multiple.' Whereas digitization adds yet another layer of representation to the historical object, it attends to the material object in terms of care, in that it enables a rare instance of 'full revision' of its fragile state as far as the whole corpus is concerned, and a last resort against the ultimate decay (disintegration). Thus, digitization is shown to be transformative not only in turning an 'analog' book into a 'digital' reproduction, but also in shaping physical objects, their representations within the library, and the divergent practices in which they are enacted. Ultimately, digitization (in the narrow sense of digitizing a particular set of objects) is an endeavor to come to terms with digitization (in the broader sense of transformation of socio-material-cultural arrangements) within this setting. The *Austrian Books Online* project is a mode of 'research and development' toward the library of an imagined future conducted through the practices of librarianship itself. This study can thus, on empirical grounds, add to contributions emphasizing 'research' not only being a domain of library and information studies as an academic discipline, but also as an inherent feature of librarianship as a profession (Cornelius 1997; see also Carlin 2009).

The ways in which this is the case have been demonstrated by trying to point out how 'multiplicity' is of intrinsic relevance within the field itself, as a members' concern. Following librarians through digitization work has shown them to be highly conscious of the 'multiple' of their objects within and beyond the library, the coordination work required to align them in day-to-day work, and the actor-networks to be stabilized so that a large-scale digitization project can be assembled successfully. As librarians explore the nature of their profession and describe their work, questions of identity and difference of objects, and socio-material relations within the setting are being raised as such questions are important to the organization of work right now and toward shifting socio-material configurations. As the analysis has shown, matters of 'ontology' are relevant within this particular setting as topics—librarianship can be said to be characterized by members' distinctive concern for questions precisely such as this. That the 'doing' of an ethnography has laid open convergences of such vernacular theorizing of conduct and their manifestation in concrete work practices, along with the conceptualizations of the analyst in the study of librarians' work, might thus be consequential to approaches adopting a general ontology of multiplicity in their explanation of objects and the social—in methodological, as well as theoretical terms.

I would thus like to propose to understand this text as an argument for shifting the basis of the study of the status of (digital) objects within librarians' practice from a distinctive ontology of multiplied realities toward empirical investigations of 'ontological topics' bound to the digital, as made relevant within such settings themselves (Lynch 2013).

Furthermore, such a stance can be valuable beyond this confined case study, in further inquiries into archival assemblages and 'the digital' within society. Adopting an 'agnostic' stance toward the 'whatness'/the difference of the digital, and instead focusing on how it is actualized in various social settings undermines the binary, quasi-ontological precondition underlying the greater part of social theory investigations of 'digital' issues.

Additionally, studying librarians' work as repair work provides an opportunity for critical engagement with digitization projects and related issues of knowledge politics beyond what is mainly discussed in public, as well as academic contexts, namely the implications of private corporations such as Google entering the domain of public knowledge infrastructures (Jeanneney 2005; Baksik 2006; Lackie 2008; Lewis 2013). As this chapter sought to elaborate, mass digitization confronts the conservation of heritage objects, and the maintenance of bibliographic order with significant challenges through the process of digitization itself. Beyond 'copies' of culturally significant works now being stored on servers in Mountain View and processed by Google's algorithms, digitization projects have profound internal effects on the maintenance and *Gestalt* of public knowledge infrastructures. Understanding these effects is critical for future developments in regulatory frameworks and policy incentives directed toward the digitization of cultural heritage (see Niggemann et al. 2011). Studying librarians' work ethnographically provides an opportunity for social theory wherein, by "letting the members teach [you]" (Randall and Sharrock 2011: 16), a critical engagement for the analyst can be based on descriptions of 'criticism' and 'critique' already internal to the field and its members' practices—by first turning 'critique' into an empirical topic on its own (Lynch 1982), and then developing a 'critical stance' on that basis.

Notes

1. Microform imaging and other reproduction technologies were part of the world of libraries throughout the twentieth century, and remain so today, where such analog storage formats still have their place in the long-term preservation of artifacts considered to be cultural heritage.

The fact that reproduction now predominantly includes transformation to digital data is, in this sense, a mere 'consequential' step, not a revolutionary 'new' endeavor. Additionally, many of the benefits and expectations generally ascribed to the transformation of documentary artifacts into digital data objects have been the pursuit of librarians and archivists (both in theory and practice) for decades, even centuries before the advent of integrated circuits and the digital computer. The aspiration to create a universal library, something that resonates in the descriptions of today's mass digitization projects, is probably over 2000 years old, as famously symbolized by the grand library of Alexandria (Thiem 1995). The possibility to provide access to 'all the world's knowledge' for enjoyment from the comfort of one's home was an ambition voiced by early information scientists such as Paul Otlet in 1910 and resulted in the establishment of an institution dedicated to this task, the *Mundaneum* (Levie 2006). In 1927, Russian-German engineer Emanuel Goldberg, at the time director of *Zeiss Ikon*, a subsidiary of German optics company *Carl Zeiss*, patented a 'work desk' allowing the retrieval of specific information from vast amounts of books and documents stored within it—a *search*, purely based on 'analog' microfilm technology (Buckland 2006). The so-called *statistische Maschine* consisted of a modified film projector and a photoelectric cell which allowed it to recognize code patterns that had been included with every previously photographed document. Not coincidentally, the principles of 'optical character recognition' (OCR) developed later, resemble this automated 'optical code recognition' system. While two prototypes had supposedly been built (Goldberg himself is said to have used one of them in his personal office), the project came to an end when Goldberg was forced to resign from *Zeiss Ikon* by the Nazi regime and had to leave Germany. The perhaps more prominent version of this concept, though only existing in the form of a magazine article, is the *Memory Extender* ('Memex') envisioned by Vannevar (Bush 1945), a work often cited as one of the most influential texts in computing history (Nyce and Kahn 1991).

2. In the Austrian National Library, the preservation of documentary heritage was extended to include digital materials in/with the year 2000. Since then, electronic media have been subjected to both legal deposit as well as collection efforts, and the library has been engaging in activities related to 'born digital resources' (i.e., materials that originated in digital form), as well as digitization. Within the context of many European

libraries' efforts in digitizing their holdings, or parts thereof, the library's management is envisioning a future wherein the complete holdings will be available in digital form by 2025, and several smaller and larger scale initiatives have been conducted already. Among them are 'on-demand' user-requested digitizations, and medium-scale digitization projects of parts of the holdings such as audio-materials or papyri. More recently, the library has also started to engage in mass digitization of historical Austrian newspapers, legislative texts, and its picture archive. One of the current main efforts in this regard is *Austrian Books Online*, a project wherein the library cooperates with Google as 'Library Partner' of the 'Google Library Program' (see Kaiser 2012). Within the project, the complete historical archive of books from the beginning of the sixteenth to the second half of the nineteenth century is being scanned and made available online, both through the library's own website, as well as *Google Book Search*. The dimensions of this undertaking are substantial, since around 600,000 volumes have to be prepared, digitally indexed, transported to an off-site scanning facility, scanned, digitally transmitted, and stored. The Austrian National Library is cooperating with Google, Inc. on this project in financial as well as operational terms. While the preparation of the books in conservation terms (the provision and management of meta-data, quality control of the scanned material, the implementation of digital long-term preservation solutions, and the development of user-access solutions) is financed and conducted by the library, the transporting of the books, the actual scanning process, image and OCR post-processing, as well as insurance costs are the responsibility of the company.

3. The most prominent example for such diagnostic approaches is perhaps Annemarie Mol's book *The Body Multiple: Ontology in Medical Practice* (2002). In this text, Mol establishes the question 'What is atherosclerosis?' as a practical question of a diverse range of "doings" in different parts of a Dutch hospital. The various practices relating to atherosclerosis each perform or "enact" a different version of this disease, "different atherosclerosis." Taking up Latour's frame toward the subject/object distinction (Latour 1993), and within the customs of actor-network theory and its semiotic cousins, she mobilizes the ethnography to propose a general ontology. Out of her empirical accounts, a discussion of constructivist social theory approaches toward the 'objective' and 'reality,' and the Goffmanian (1990 [1959]) conception of *performance* of

self, Mol develops a central philosophical assertion. The enactment of an 'objective' entity through socio-material practices reveals the status of the object as *multiple*: "[o]ntology-in-practice is multiple" (Mol 2002: 157).
4. The entry reproduced here resembles the simplified version an end-user would see when searching the online catalog. The original can be accessed at http://data.onb.ac.at/rec/AC10286864 (Retrieved 15 September 2015).

References

Baksik, Corinna. 2006. Fair use or exploitation? The Google Book Search controversy. *Portal: Libraries and the Academy* 6 (4): 399–415.

Bishop, Ann Peterson, Nancy A. Van House, and Barbara P. Buttenfield (eds.). 2003. *Digital Library Use: Social Practice in Design and Evaluation.* Cambridge: MIT Press.

Buckland, Michael K. 1997. What is a 'document'? *Journal of the American Society for Information Science* 48 (9): 804–809.

———. 2006. *Emanuel Goldberg and His Knowledge Machine: Information, Invention, and Political Forces*. Westport, CO: Libraries Unlimited.

Bush, Vannevar. 1945. As we may think. *The Atlantic Monthly* 176 (1): 101–108.

Carlin, Andrew P. 2009. Segmented professions: Further considerations of theory and practice in LIS and librarianship. *Library Philosophy and Practice*. October. Retrieved 10 March 2014 http://unllib.unl.edu/LPP/carlin.pdf.

Ceruzzi, Paul E. 2012. *Computing: A Concise History*. Cambridge: MIT Press.

Cornelius, Ian. 1997. Research as practice. In *Library and Information Studies: Research and Professional Practice*, ed. Micheline Beaulieu, Elisabeth Davenport, and Nils Ole Pors, 243–248. London: Taylor Graham.

———. 2004. Information and its philosophy. *Library Trends* 52 (3): 377–386.

Coyle, Karen. 2006a. Mass digitization of books. *The Journal of Academic Librarianship* 32 (6): 641–645.

———. 2006b. One word: Digital. *The Journal of Academic Librarianship* 32 (2): 205–207.

Cubitt, Sean. 2006. Library. *Theory, Culture & Society* 23 (2–3): 581–590.

Denis, Jérôme, and David Pontille. 2015. Material ordering and the care of things. *Science, Technology and Human Values* 40 (3): 338–367.

Ernst, Wolfgang. 2002. *Das Rumoren der Archive: Ordnung aus Unordnung*. Berlin: Merve Verlag.

Evens, Tom, and Laurence Hauttekeete. 2011. Challenges of digital preservation for cultural heritage institutions. *Journal of Librarianship and Information Science* 43 (3): 157–165.

Gad, Christopher, and Casper Bruun Jensen. 2009. On the consequences of Post-ANT. *Science, Technology and Human Values* 35 (1): 55–80.

Garfinkel, Harold, and Harvey Sacks. 1986 [1970]. On formal structures of practical actions. In *Ethnomethodological Studies of Work*, ed. Harold Garfinkel, 160–193. London and New York City, NY: Routledge & Kegan Paul.

Goffman, Erving. 1990 [1959]. *The Presentation of Self in Everyday Life*. London: Penguin Books.

Goodwin, Charles. 1994. Professional vision. *American Anthropologist* 96 (3): 606–633.

Graham, Stephen, and Nigel Thrift. 2007. Out of order: Understanding repair and maintenance. *Theory, Culture & Society* 24 (3): 1–25.

Gregson, Nicky, Alan Metcalfe, and Louise Crewe. 2009. Practices of object maintenance and repair: How consumers attend to consumer objects within the home. *Journal of Consumer Culture* 9 (2): 248–272.

Henke, Christopher R. 1999. The mechanics of workplace order: Toward a sociology of repair. *Berkeley Journal of Sociology* 44: 55–81.

Ingold, Tim. 2007. Materials against materiality. *Archaeological Dialogues* 14 (1): 1–16.

Jackson, Steven J. 2014. Rethinking repair. In *Media Technologies: Essays on Communication, Materiality, and Society*, ed. Tarleton Gillespie, Pablo J. Boczkowski, and Kirsten A. Foot, 221–293. Cambridge: MIT Press.

Jeanneney, Jean-Noël. 2005. *Quand Google défie l'Europe. Plaidoyer pour un sursaut*. Paris: Mille et une nuits.

Jones, Siân, and Thomas Yarrow. 2013. Crafting authenticity: An ethnography of conservation practice. *Journal of Material Culture* 18 (1): 3–26.

Kaiser, Max. 2012. Putting 600,000 books online: The large-scale digitisation partnership between the Austrian National Library and Google. *Liber Quarterly* 21 (2): 213–225.

Knorr-Cetina, Karin D. 1999. *Epistemic Cultures: How the Sciences Make Knowledge*. Cambridge, MA: Harvard University Press.

Lackie, Robert J. 2008. From Google Print to Google Book Search: The controversial initiative and its impact on other remarkable digitization projects. *The Reference Librarian* 49 (1): 35–53.
Latour, Bruno. 1986. Visualisation and cognition: Drawing things together. In *Knowledge and Society: Studies in the Sociology of Culture Past and Present*, vol. 6, ed. Henrika Kuklick and Elizabeth Long, 1–40. Greenwich, CT: Jai Press.
———. 1993. *We Have Never Been Modern*. Cambridge, MA: Harvard University Press.
Latour, Bruno, and Adam Lowe. 2011. The migration of the aura, or how to explore the original through its facsimiles. In *Switching Codes: Thinking Through Digital Technology in the Humanities and the Arts*, ed. Thomas Bartscherer and Roderick Coover, 275–297. Chicago, IL: University of Chicago Press.
Law, John, and John Hassard (eds.). 1999. *Actor Network Theory and After*. Oxford and Malden, MA: Blackwell.
Levie, Françoise. 2006. *L'homme qui voulait classer le monde. Paul Otlet et le Mundaneum*. Bruxelles: Les Impressions Nouvelles.
Levy, David M. 2003. Documents and libraries: A sociotechnical perspective. In *Digital Library Use: Social Practice in Design and Evaluation*, ed. Ann Peterson Bishop, Nancy A. Van House, and Barbara P. Buttenfield, 25–42. Cambridge: MIT Press.
Lewis, Ben. 2013. *Google and the World Brain*. Polar Star Films.
Lynch, Michael E. 1982. Technical work and critical inquiry: Investigations in a scientific laboratory. *Social Studies of Science* 12 (4): 499–533.
———. 2013. Ontography: Investigating the production of things, deflating ontology. *Social Studies of Science* 43 (3): 444–462.
Mol, Annemarie. 1999. Ontological politics: A word and some questions. In *Actor Network Theory and After*, ed. John Law and John Hassard, 74–89. Oxford and Malden, MA: Blackwell.
———. 2002. *The Body Multiple: Ontology in Medical Practice*. Durham and London: Duke University Press.
Nichols, Stephen G. 2006. An artifact by any other name: Digital surrogates of medieval manuscripts. In *Archives, Documentation, and Institutions of Social Memory: Essays from the Sawyer Seminar*, ed. Francis Xavier Blouin Jr. and William G. Rosenberg, 134–143. Ann Arbor: University of Michigan Press.
Niggemann, Elisabeth, Maurice Lévy, and Jacques De Decker. 2011. The new Renaissance: Report of the comité des sages on bringing Europe's cultural

heritage online. Retrieved 4 October 2013 http://ec.europa.eu/information_society/activities/digital_libraries/doc/refgroup/final_report_cds.pdf.

Nyce, James M., and Paul Kahn (eds.). 1991. *From Memex to Hypertext: Vannevar Bush and the Mind's Machine*. Boston, MA: Academic Press.

Randall, Dave, and Wes W. Sharrock. 2011. The sociologist as movie critic. In *Ethnomethodology at Work*, ed. Mark Rouncefield and Peter Tolmie, 1–18. Farnham: Ashgate.

Rieder, Bernhard. 2013. 81.498 words: The book as data object. In *The Unbound Book*, ed. Adriaan van der Weel and Joost Kircz, 57–70. Amsterdam: University Press.

Schütz, Alfred. 1971 [1953]. Wissenschaftliche Interpretation und Alltagsverständnis menschlichen Handelns. In *Gesammelte Aufsätze I: Das Problem der sozialen Wirklichkeit*, 3–54. Den Haag: Martinus Nijhoff.

Thiem, Jon. 1995. Myths of the universal library: From Alexandria to the postmodern age. *The Serials Librarian* 26 (1): 63–74.

Watson, Rod, and Andrew P. Carlin. 2012. 'Information': Praxeological considerations. *Human Studies* 35 (2): 327–345.

Westin, Jonathan. 2013. Loss of culture: New media forms and the translation from analogue to digital books. *Convergence: The International Journal of Research into New Media Technologies* 19 (2): 129–140.

Woolgar, Steve, and Javier Lezaun. 2013. The wrong bin bag: A turn to ontology in science and technology studies? *Social Studies of Science* 43 (3): 321–340.

Zimmerman, Don H., and Melvin Pollner. 1970. The everyday world as a phenomenon. In *Understanding Everyday Life: Toward the Reconstruction of Sociological Knowledge*, ed. Jack D. Douglas, 80–103. Chicago, IL: Aldine Publishing.

4

Job Done: What Repair Does to Caretakers, Tenants and Their Flats

Alain Bovet and Ignaz Strebel

Introduction

Tenants live, and sometimes spend their whole lives, in flats they do not own. When something breaks down in the material infrastructure of the flat, they are entitled to get it repaired by the owner, who puts in place the necessary procedure to process repair work. In the case investigated for this paper, tenants call a caretaker who then appraises whether his[1] intervention or the intervention of a specialist is needed. In some cases, the caretaker can guide the tenant to solve the problem him- or herself. If the caretaker decides to intervene himself, he will see the tenant at his

A. Bovet (✉)
HEG—Haute école de gestion Arc, University of Applied Sciences and Arts Western Switzerland, Neuchâtel, Switzerland
e-mail: alain.bovet@he-arc.ch

I. Strebel · A. Bovet
Institute of Geography and Sustainability, University of Lausanne, Lausanne, Switzerland
e-mail: ignaz.strebel@unil.ch

I. Strebel et al. (eds.), *Repair Work Ethnographies*,
https://doi.org/10.1007/978-981-13-2110-8_4

or her flat and attempt to solve the problem. Unless the tenant appears directly responsible for the damage, he or she will not be charged for the repair (which is taken to be covered by the rent and/or additional costs they pay to the landlord). While formally the caretaker repairs broken objects that are not owned by the tenant, they have to take into account that they enter and then work at the request of the inhabitant and in his or her private sphere. The question that is raised by the troublesome flat is then not so much "who owns what?" than "who is in charge of what?"

This chapter aims to show how these aspects specify the kind of repair work that is accomplished by caretakers in rented flats. From a praxeological perspective, we will attempt to demonstrate that caretaker interventions display an orientation to the fact that they repair things at the request, and in the presence of one or several persons who, while they may not own these repairable things, do live with them. Our focus is on the end of the job and the *closing* of interventions, which are particularly illustrative of this orientation and a crucial phase for making salient the various dimensions of caretaker repair work. The *closing* is a sequence that reconfigures the relations among caretaker, broken thing, and tenant. Successful or not, the repair work never brings back the initial (i.e., pre-breakdown) configuration. Once the repair has come to a close, the caretaker and the tenants have learned something: the thing has changed and its future use by the tenant will not be the same. While all these aspects emerge in the course of the repair work, they are especially at stake when the intervention of the caretaker is brought to its conclusion.

Although much of the closing is accomplished through talk between the caretaker and tenants, it cannot be reduced to an exchange of words. The talk is not just about material things, it is embedded in and constrained by a material ecology, which is not only comprised of the repaired thing, but also the whole flat, as well as the bodies of the interactants, tools, spare parts, etc. For this reason, our analyses will be based on excerpts of video recordings of caretaker interventions. The video data have been gathered in an ethnographic study on building care in residential buildings and on housing estates in Switzerland.

The video ethnography involved documenting the working routines of professional caretakers of a housing administration, a housing cooperation, one mobile facility-service company, and a real-estate family business located in the greater urban areas of Zurich, Basel, and St. Gallen in the Swiss Midlands. One member of our research team, accompanied by a filmmaker, documented a full-work day and various tasks performed by the five research participants. Additionally, all participating caretakers, their employers, and a selected number of colleagues were interviewed. The shadowing technique "following the concierge at work" (Jacobs et al. 2012) aimed at making visible the ordinariness and in situ accomplishment of caretaker work, its routine patterns, its contingencies, and its emergency dimensions, while highlighting technical and social problems that maintenance staff in residential buildings are faced with. The twenty-four hours of video and audio recordings collected over a one-year period were logged and transferred into a searchable data corpus. The main goal of the data corpus was to collect video-recorded caretaker practices for analysis of multimodal interaction. As well as this analytic concern, and as a way of offering feedback to research participants, selected video footage was also assembled into a documentary film, which pays homage to the day-to-day work of building care in housing, and gives a first-hand insight into the richness of the phenomena found in the data.[2]

This video ethnography allowed us to take a renewed interest in residential buildings "after they're built" (Brand 1994) with a specific focus—unusual in housing studies—for once not on the resident, but on those people who work with residential buildings and their populations. In doing so, the video ethnography was set up in the tradition of ethnomethodological workplace studies (Garfinkel 1986; Luff et al. 2000), inspired by studies of occupations (Gamst 1980; Harper 1987), and the few available studies on the work of caretakers (Gold 1952, 1964), *gardiens* (Stébé and Sudant 2002; Laé 2015; Marchal 2006; Ughetto 2011), doormen (Bearman 2005), and *Hausmeister* (Omahna 2003; Payer 1996) in urban contexts. Recent contributions have reconsidered caretaker work as a specific social relation of care, in which service agents and technicians engage with users, not according to given rules or professional positions, but with the aim of understanding their ability to care for and respond in situ to problematic situations

(Bonnet 2014). For this chapter, also of interest are those ANT-driven housing studies that have emphasized the ways in which maintenance routines are organized and how such routines contribute to shaping buildings and infrastructure as multiple achievements (Edensor 2011; Jacobs and Cairns 2011). Yet, while studies on care work as a social relationship are not concerned with relocating the materiality of buildings and building technology in the interactions between housing agencies, caretakers, and tenants, socio-technical studies written in a symmetrical mode leave unanswered the praxeological concern about how the realities described were practically achieved as such: How do caretakers carry out diagnostic work so that they can identify and prepare the tools to bring to the workplace? How would caretakers respond to and deal with tenants' ad hoc evaluations of the problem? How do they guide and advise colleagues and building craftsmen? What are their suggestions for and formulations of an appropriate solution? While such questions remain unstudied, the video-documentary approach offers us an opportunity to move from the assumption *that* a viable building is a multiple achievement, to the explanation of *how* such an achievement is realized.

It is in this sense that we shift our attention in this chapter to the moment in which repair work in the tenant's flat comes to a close, arguing that it is at this very moment that the orientation to viability is clearly displayed. While this is heuristically useful for our analysis, we also want to show that it gives another insight—beyond the open/closed, stable/unstable black box—into the conceptual thinking of repair work as a component of infrastructure. On the way, by analyzing selected video clips of such closing moments in our data corpus, we will move from a clear-cut understanding of what is opened and what is closed, or of what is broken and what is repaired, to a possibly less clearly traceable but integrated role of breakdown and repair, when it comes to maintaining the viability and usability of residential buildings. In the first part of the chapter, we will demonstrate how repair work is oriented to the viability of the flat. In the second part, we will elaborate on the way repair work changes this viability beyond the repair situation. Finally, we will discuss what can be learned about repair work from this specific moment of closing.

Orientations of Repair

When a caretaker comes to a flat to do repair work, it is not just to talk with the tenant, which he could do on the phone. He is expected instead to operate on a material thing that is broken. However, his intervention can neither be reduced to technical matters alone, nor can the response to a call for a technical operation be reduced to the social context of the intervention. Call and response are rather an integral part of the intervention. In other words, the caretaker does not just repair but does so for the tenant; more precisely, his work is oriented[3] to the viability of the flat.[4] We will argue that this is what makes this course of action *accountable* as a caretaker intervention, following Garfinkel's seminal ethnomethodological intuition that "the activities whereby members produce and manage settings of organized everyday affairs are identical with members' procedures for making those settings 'accountable'" (Garfinkel 1967: 1). We will now turn to a first example showing that the social and the technical aspects of the intervention may be treated as distinct phases.

Two Clearly Distinguishable Finishing Phases: "Finito" vs. "So It Is Finished uh?"

In the following case, the caretaker is called by a tenant about his kitchen tap. While he can still use the tap, there is too much play when he moves it to the extent that he is scared of breaking it. For the caretaker, who has done this job many times in other flats, it is a routine operation consisting of replacing the inner part of the tap.

Once the caretaker has replaced the damaged part, he reassembles the other components of the tap. Once the last component is installed, he tries the tap (Fig. 4.1) and mumbles to himself (and audible to the cameraperson) "so, *finito*" ("so, finished"), in the typical form of self-talk that accompanies much caretaker work. He then grabs a paper tissue and begins rubbing the surface of the tap. He later engages in a variety of tasks such as reopening the water valve in the cupboard below the sink, cleaning up, replacing the tools in the toolbox, etc.

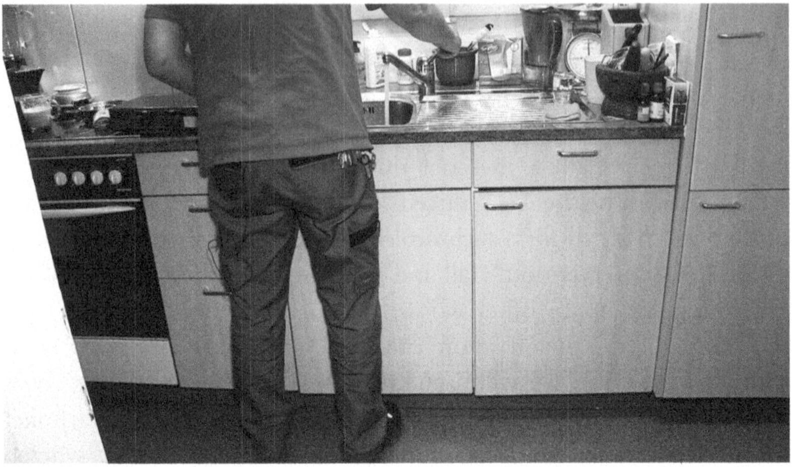

Fig. 4.1 Trial for oneself (© Video still: I. Strebel)

The sequential placement of "*finito*" indicates neither the departure of the caretaker, nor even the completion of the intervention. We will even see that there is another distinct way of telling the tenant that the repair is finished. What "*finito*" signals is apparently the completion of the strictly technical intervention on the tap. Though the task left to complete at this point is about as long as what has already been done, it is characterized as a distinct type of work.

Four minutes later, an interaction with the tenant takes place. At the beginning of the excerpt, the tenant is in another room.[5] The following transcript proposes a textual representation of the excerpt and is complemented with video stills. Non-verbal elements are described in double parenthesis. Verbal elements are transcribed as follows: the first line displays the abbreviated name of the speaker (here "Car" for the caretaker and "Ten" for the tenant) and the original talk in Swiss German. The second line, in bold italics, is an English translation. Dots stand for short pauses. Longer pauses are numbered in seconds in simple parentheses. The line numbers will be used in the following analysis to index specific phenomena.

Transcript 1

((The caretaker closes the cupboard below the sink and	
turns on the tap in every position of the lever.))	
Car	so isch fertig hä
	so it is finished uh
	(4.0)
((He turns off the tap, puts his toolbox on the	
floor and rubs his hands. The tenant approaches the scene.))	
Ten	scho
	already
Car	JÄ . äh wiä noi . hä?
	YES . as new . uh?
((The caretaker turns on the tap in every position	
of the lever. When the tenant reaches the tap, the	
caretaker lets go of the lever.))	

Just before the beginning of the excerpt, the caretaker completes the last operation on the water valve below the sink. He then closes the cupboard and tries the repaired tap in many positions. The trial being successful, he formulates the completion of the repair (line 3). This piece of talk is obviously voiced differently from the mumbled "*finito*" described earlier. As a summons (Schegloff 1968), it is the first part of an adjacency pair (Schegloff and Sacks 1973), the second part of which is to be fulfilled by the tenant.[6] It is then an announcement to the tenant of the—obviously successful—achievement of the repair.

The tenant's reply (8) expresses mild surprise at the quick achievement of the task. This is quickly and emphatically confirmed by the caretaker (10), who upgrades the assessment with a qualification of the repair, which makes the present tap not just repaired, but as good as new. Anticipating the tenant's arrival at the tap, the caretaker displays that it is working in each position of the lever and then let's go of it (12–14).

This action is oriented to the tenant in at least two respects. First, it displays that the lever is no longer loose, which was what had worried

Fig. 4.2 Display of the repaired thing to the tenant (© Video still: I. Strebel)

the tenant and made him call the caretaker. Second, it is clearly produced not as a trial, like the one the caretaker carried out just before uttering the summons (2), but so as to be optimally visible to the approaching tenant, as can be seen on the video still (Fig. 4.2).

What the caretaker does is actually more than a display in that it invites the tenant to try the new tap. We have seen so far that the caretaker produces two clearly distinct phases of the conclusion of his repair work. The first utterance ("so, *finito*") seems mostly addressed to himself as a way of structuring the ongoing work. The second is addressed and oriented to the tenant. If the tenant had not answered the summons, it would have produced a remarkable absence. As we suggested earlier, this intersubjective phase is no more and no less a part of the intervention than the technical operation on the tap. In other words, the technical operation only makes sense as the object of a tenant's request to a caretaker and thus, has to be made accountable to him. As we will see in the second section, this phase is not only devoted to producing an intersubjective achievement of the repair, but it also opens a sequence in which aspects of the job processed and the repaired object are revealed both discursively and practically to the tenant, in particular, in terms of implications for the future use of the tap.

The case we have just examined is helpful in identifying and distinguishing two phases of repair work and in particular, its closing. We now turn to a case where those two phases are merged in a very short timeframe. We will attempt to show that such a contraction requires a specific action by the caretaker in order to make his intervention accountable to the tenant.

Intertwined Finishing Phases—"Oh Now It Works Again" Followed by "the Key Was Not All the Way in"

A tenant comes to another caretaker's office with a question: "Do we not have power in the laundry room?" The caretaker is busy talking on the phone, but interrupts his call briefly and promises the tenant that he will "come and have a look right away". A few minutes later, the caretaker enters the laundry room and sees that the washing machine has no power. With the unsolicited help of the tenant, he then engages in an inquiry into the cause of the problem. It is during the course of this inquiry that the problem is suddenly solved (Fig. 4.3).

In the following transcript, utterances delimited with "°" are said in a distinctly low voice, almost whispered.

<u>Transcript 2</u>

Car ja dänn muäs i gschwind go luägä
°wart villicht hät är do no öppä°
yes then I must quickly go have a look
°wait maybe has it here possibly°
((The caretaker turns the key off and on and gives it
two little pushes.))
Car °da muäs i gschwind go luägä°
°there I must quickly go have a look°
((The caretaker returns to the tenant while looking at the
machine then stops.))
Car ah luäg jetzt
oh look now

Ten ah jetzt gots widär
oh now it's working again
((The caretaker moves back towards the key.))
Car dr schlüssäl nit ganz din gsi
the key was not all the way in

Fig. 4.3 Course of inquiry and problem solved (© Video still: I. Strebel)

The inquiry the caretaker is engaged in consists of looking at and manipulating various parts of the machine and its power supply. It is not a strict trial and error sequence in which the caretaker would perform a single action and then check whether the machine works again, and if not, try something else. At the beginning of the excerpt, he manipulates the power supply key (1–6). He then has another idea, probably to check the fuse in another room. On his way, he nevertheless takes a peek at the machine and notices by chance that it is working again (9–10). He stops and points it out to the tenant with a typical change of state token (11). The tenant confirms it with a similar utterance (13). The caretaker then goes straight back to the key and provides a discursive formulation of the source of the trouble. Repair apparently comes to a close here. However, as we will see later, what constitutes a close must be examined carefully. According to our praxeological perspective, an action only comes to a close when accomplished and made accountable as such by the participants.

The excerpt we have just examined (*"oh now it works again" followed by "the key was not all the way in"*) exhibits a case of a repair that is distinct from the previous one ("finito" *vs. "so it is finished, huh?"*). Here the inquiry into the source of the trouble and the repair are simultaneous. Repair could even be said to precede the discovery of the source of the trouble. The fact that the machine works again triggers a retrospective inquiry into the manipulation that may have caused it. It is immediately clear to the caretaker that the manipulation of the key is both the source of the trouble, as well as the means to repair it. While this may correspond to the "*finito*" examined before, what remains is to make clear to the tenant that the repair has been achieved, and how this has been done. This second part of the job is immediately accomplished by the caretaker, who returns to the "crime scene" (15), and provides a discursive formulation of the source of the trouble (16). This formulation is, of course, also meant to be understood as an instruction to the tenant on how to manipulate the key.

This example shows that providing for the intelligibility of the repair to the tenant is an internal requirement of the caretaker intervention.

It is part of the professional ethics of the caretaker, to the extent that a caretaker who would stick to the technical job and abstain from making it intelligible to the tenant, might be an excellent technician, but would not be a caretaker. In the case just examined, the accountability is provided immediately after the technical repair job has been finished. What is particular and surprising, but nevertheless pertinent in this sequence, is that the repair action is identified retrospectively and inductively on the basis of the consequence of manipulation of the machine.

Our point in this first section of the chapter is not to contest the distinction between the technical operation and how it is made accountable, i.e., intelligible and acceptable to the tenant. Rather, we have attempted to show that this distinction is relevant to the participants and has consequences for the course of the caretaker's intervention.[7] In the case just examined, the tenant is present the whole time. He cannot then call her once the repair is achieved and ready to be displayed. He must therefore organize the repair accountability on the spot, within the co-presence situation.

In the two cases we have examined so far, the repair is successful and is therefore announced as such to the tenant. To conclude the first part of the chapter, we now turn to an unsuccessful repair. We will see that repair failure produces a different configuration of the closing of the intervention and as such assigns a special task to the caretaker.

Failed Finishing Phase—"I Don't Remove the Heating Rod, Now I Have to Figure Something Else Out"

The unavoidable fact that repair does sometimes fail does not mean that repair failure constitutes an acceptable option. Just as for the photocopy machine technicians observed by pioneer repair ethnographer Julian Orr (1996), repair failure is not an acceptable outcome of the caretaker intervention. A repair attempt may fail but the caretaker remains in charge of the trouble and has to explain what this entails for the tenant. This peculiar horizon of caretaker repair work is nicely illustrated in the following scene, which also shows that, successful or not, repair is oriented to the viability of the flat.

In the following case, the caretaker has been called by tenants for a breakdown of the fan-assisted function in the oven. On the basis of the call, the caretaker considers the situation as a common case of heating rod failure, which can be routinely fixed by replacing the fan function heating rod.[8] But the actual repair process proves much more problematic. After removing the oven door and the back plate inside the oven, the caretaker discovers that the heating rod's screws are completely rusted. His successive attempts at, first, unscrewing it with a screwdriver; second, removing it with a drill; and finally, a hammer, all prove unsuccessful. This produces a kind of reverse scenario to the "*finito*" case we examined before: the caretaker reaches a point where he knows that he will not be able to unscrew the heating rod and finish the repair. He first formulates it in the following joke that is addressed to the cameraperson: "Let's put it back together and act as if there is nothing wrong". He then addresses the problem with the tenant in a much more serious and responsible manner. The following exchange takes place after the caretaker has finished restoring things to their original state and the tenant has entered the kitchen (Fig. 4.4).

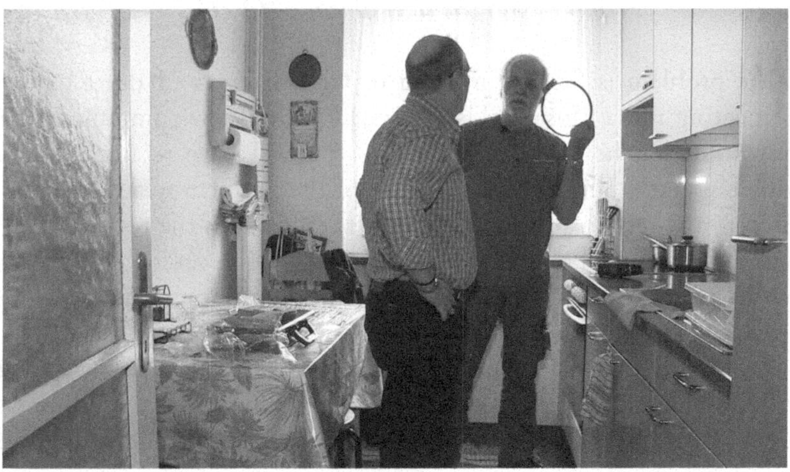

Fig. 4.4 The heating rod is the source of the problem (© Video still: I. Strebel)

Transcript 3[9]

1	Car	ich bring d heizig nöd äwäg
2		jetzt muäs i miär öppis ifallä la
3		***I don't remove the heating rod***
4		***now I have to figure something else out***
5	Ten	welche heizig?
6		***which heating rod?***
7	Car	diä wo kaputt isch
8		***the one that is broken***
9	Ten	kanns nit weg?
10		***can't it be removed?***
11	Car	chann nit wägnä isch total verroschtet und
12		***can't remove it it's completely rusty and***
13	Ten	und was machen jetzt?
14		***and what to do now?***
15	Car	jetzt muäs i öppis studiärä
16		oder montör cho la ich tuä luägä
17		***now I have to figure something else out***
18		***or have a technician come I'll see***
19	Ten	oder eine neue und
20		***or a new one and***
21	Car	ja das hät si scho gsait HA HA HA HA HA HA
22		***yes that's what she said already HA HA HA HA HA HA***

The problem is first formulated in factual terms: "I don't remove the heating rod". This turn is expanded with a new component that addresses what he intends to do next: "now I have to figure something else out". This first formulation of the problem seems to be oriented to the responsibility of the caretaker vis-à-vis the tenant, rather than an account of the failure of the repair. The caretaker indicates that he remains in charge of the problem, even though the repair has failed.

The tenant then asks the question "which heating rod?" (6), which expresses ignorance of the heating rod that the caretaker just mentioned. The latter's brief answer "the one that is broken" (8) does not engage in a description of the various parts of the oven. Instead, he simply specifies the heating rod in terms of the repair process as the source of the trouble, making it clear that the repair has failed. The tenant then asks a second question as a correction-invitation device ("can't it be removed?" (10)), which reformulates the initial description of the problem. It should, however, be noted that the reformulation introduces the verb "can" which, contrary to the initial factual formulation by the caretaker, addresses the ability or inability to remove the heating rod. This is maintained in the caretaker's answer "can't remove it it's completely rusty" (12), which begins as an admission of inability, but is immediately followed by an explanation, which exempts the caretaker from any responsibility.[10] The rust is a state of the matter that is discovered and encountered, especially as it prevents any planned action on the rusted material. We will return to this later.

The tenant does not further explore the cause of the repair failure. He even interrupts the caretaker's developing turn to ask what the next step will be and thus appears to share the caretaker's central concern about the incompletion of the repair process. The caretaker's reply, "now I have to figure something else out or have a technician come, I'll see" (17–18), starts with a reformulation of a previous utterance that states that he is required to find a solution. The turn is completed with an indication that makes clear that *he* will make the decision, stressing again that he is, and continues to be, in charge of the repair. In other words, even if it is not yet clear what will be done, something will be decided and done by the caretaker.

In spite of this, the tenant offers another option, suggesting replacement of the oven with a new one. The caretaker immediately indicates that the tenant's wife has already offered that suggestion and bursts into loud laughter. At the beginning of the whole intervention, the caretaker had rejected the tenant's wife's replacement proposal via the rule-like argument that a 25-year-old oven could be replaced, whereas the 15-year-old one in question must be repaired.

Both parts of the caretaker's reply ("yes that's what she said already HA HA HA HA HA HA HA HA") make clear that the tenant's proposal is not feasible: it has already been proposed and rejected, thus making its reiteration funny. The reply appears as a polite, or at least not overly rude, way of avoiding a rather uneasy discussion. The rust, at least in common sense reasoning, could readily be taken as a sign of aging, and the fact that it prevents the repair is but a supplementary argument in favor of its replacement. The caretaker's reply then can be said to adjust to this precarious interactional environment.

This contrasting case shows that the outcome of the intervention is a concern for both caretaker and tenant. If repair fails, the caretaker must provide a form of not only retrospective accountability through explanations and apologies,[11] but also prospective accountability by envisaging what will nevertheless be done. The caretaker thus makes clear that he remains in charge and in control of the procedure.

So far, we have shown that repair consists not only of a technical operation, but also of making it accountable to the tenant. As a consequence, the conclusion of repair is more complex than previously assumed. On the one hand, repair can be finished as a technical and organizational matter, as in the "*finito*" case. On the other hand, it is finished with the caretaker telling the tenant whether it has been successful, as in the "so it is finished huh?" and the "the key was not all the way in" cases; or unsuccessful, as in the "now I have to figure something else out" case.

Until now we have by and large assumed that there is an end and we have, while searching for this end, accounted for the many things necessary to work toward this goal. We will now show that closing goes beyond finishing the job. In the second section of this chapter, we will explore some other aspects of the intersubjective achievement and closing of accountable repair work. We will focus on what the repair does or does not change for the future use of the repaired object for the tenant and how both caretaker and tenant(s) orient to this concern.

Life After Repair

Many repair scholars have noted the internal tension between preservation and transformation (Lejeune in this volume).[12] This tension carries a special significance in caretaker intervention. On the one hand, repair is intended to restore the flat to its pre-breakdown state. On the other hand, some aspects of the flat are inevitably modified in the process, which means that some aspects relating to the viability of the flat or the tenants' life in the flat will be modified (see Gregson et al. 2009 on family life with repaired pieces of furniture). We will see in this second part of the chapter how tenants and caretakers deal with this tension. To do so, we will simply return to the cases studied in the first part of the paper to examine how the repair process reconfigures the association of the tenant and the formerly-broken-now-repaired thing. We will argue that this form of prospective accountability is an integral part of repair work.

Repaired Tap, New Use

We will now return to the repair sequence in which the caretaker had to replace the damaged inner part of a kitchen tap. We have seen that he distinguished two steps of the repair's conclusion ("*finito*" vs. "so it is finished uh?") and we analyzed the second, which was addressed to the tenant. When the tenant comes into the kitchen, the caretaker displays the now repaired tap and invites the tenant to try it. We will now examine what ensues until the point at which the caretaker says goodbye to the tenant and leaves the flat. This long sequence will be analyzed through three transcript extracts.

<u>Transcript 4 (extract 1)</u>

((The caretaker closes the cupboard below the sink and turns on the tap in each position of the lever.))	
Car	so isch fertig hä
	so it is finished uh
	(4.0)
((The caretaker turns off the tap, puts his toolbox on the floor and rubs his hands. The tenant comes in.))	
Ten	scho
	already
Car	JÄ . äh wiä noi . hä?

11		***YES . as new . uh?***
12	((The caretaker turns on the tap in each position	
13	of the lever. When the tenant reaches the tap,	
14	the caretaker lets go of the lever.))	
15	Car	er lauft jetzt halt da no chli
16		***it's still running a bit here***
17	Ten	ja unglaublich ja ja
18		***yes unbelievable yes yes***
19	((The caretaker holds the hose and turns it in	
20	both directions. The tenant turns on the tap in	
21	each position of the lever.))	
22	Car	no chli sträng wiil är äh
23		noii gümmi dinnä hät oder
24		***still a little hard because***
25		***it has new joints in***
26	((The caretaker briefly taps the hose twice and moves	
27	to the left. The tenant lets go of the lever, holds	
28	the hose, and turns it in both directions.))	

When the tenant enters the kitchen, the caretaker lets go of the lever and thus invites the tenant to manipulate it (12–14), which he does. But the caretaker then holds the hose and keeps it until he has completed a warning (15, 22–23): "It runs still a little hard because it has new joints put in". The warning is interrupted by the tenant, who provides the first positive assessment based on the trial of the lever (17). Once the warning has been given, the caretaker lets go of the hose and moves to the left (26–27), inviting the tenant to try it.

The caretaker's warning addresses here a somewhat paradoxical aspect of repair: while the breakdown is solved, the actual use of the repaired thing may not be equivalent to the pre-breakdown use. The caretaker's warning not only prevents an expectable disappointment of the tenant,[13] but also qualifies the new use in at least two respects. First, this "hard run" is provisional ("still"). Just as new shoes have to be worn to better fit the feet, the new tap will improve through use, without the need for anything extra to be done. Second, the reason the caretaker offers for the "hard run" is the replacement of the inner joints.

It provides for the accountability of the caretaker's work as a professional intervention.[14] In other words, what may appear as a defect of the repaired thing is turned into a quality, as well as an indication, of the competent work that produced it. The positive assessment of the caretaker's work continues in the rest of the sequence.

Transcript 4 (extract 2)

26	((The caretaker briefly taps the hose twice and moves	
27	to the left. The tenant lets go of the lever, holds	
28	the hose and turns it in both directions.))	
29	Ten	ah ja guät gäg vornä isch das äh
30		sehr ring xxx merci vielmal
31		***oh yes all right compared to before this is***
32		***uh very easy xxx thank you so much***
33	((The tenant lets go of the hose, holds the lever and	
34	turns on the tap in each position of the lever.))	
35	Car	bitte jetzt isch halt no chli fettig müand
36		no chli abwäschä mit soifä dass isch wägäm
37		hahnäfett hä
38		***you're welcome now it is still a little greasy***
39		***you will have to wash it with soap this is***
40		***because of the tap grease***
41	((The tenant turns off the tap.))	
42	Ten	kais problem super
43		***no problem great***

Once the tenant has tried turning the hose (27–28), he provides another positive assessment (29–30). It is formulated in terms of a comparison with the previous state, before the tap was repaired. This assessment, which is completed with thanks, may be understood as an answer to the warning, in that instead of finding that it is hard to turn compared with its pre-breakdown phase, the tenant chooses to say that it turns more easily than it did in its pre-repair, i.e., post-breakdown phase. This suggests, then, that both parties take pains to make the repair "go smoothly".

In a complex turn (35–37), the caretaker acknowledges the tenant's congratulations ("you're welcome") and provides another warning ("now it is still a little greasy"), which is followed by advice ("you will have to wash it with soap") and an explanation ("this is because of the tap grease"). The tenant replies (42) with an acceptance and another positive assessment ("no problem great").

The caretaker here addresses an aspect of the repaired tap that is neither equivalent to the "pre-breakdown" nor to the "pre-repair" phase. While not related to the problem itself, it appears as a collateral effect of the repair. The advice is followed by an instruction to the tenant on how to deal with it, by which the caretaker indicates that this part of the repair work belongs to the tenant. Yet the caretaker offers an explanation that relates the current state of the tap to the technical operation that he has just completed. The tenant then clearly ratifies the caretaker's proposed division of labor. This exchange indicates that the caretaker is not in charge of every task by which the flat is made, if not exactly as before, at least as livable and viable as possible. Some, but not all, aspects of cleaning belong to the tenant. The same holds for putting things back as they were, as is shown in the next extract.

<u>Transcript 4 (extract 3)</u>

44	((The tenant turns on the tap. He moves the lever	
45	in both directions. The caretaker re-installs the	
46	mat beneath the sink.))	
47	Car	ah . jetzt chönn mr das
48		***uh. now this we can do***
49	Ten	ja vorig häts mir angst gmacht hä
50		***yes before that it scared me***
51	Car	nai angst müand är kei ha
52		***no you do not have to be scared***
53	Ten	nei nur sii loo
54		***no you can leave it***
55	((The tenant turns the tap on and off.))	
56	Car	äs cha nit meh als kaputt ga ha ha ha ha .

57 so tragisch isch das nöd
58 ***it cannot do more than break ha ha ha ha .***
59 ***it is not that tragic***
60 ((The caretaker re-installs the chopping board next to
61 the sink. The tenant turns towards the camera.))
62 Ten zruggbaut wird au no ha ha
63 ***it's going to be restored yet ha ha***
64 Car okay
65 ***okay***
66 Ten merci herr gunzähuser
67 ***thanks mister gunzenhauser***
68 Car also in däm fall än schönä ade
69 ***so in that case have a nice one***
70 ((The caretaker and the tenant shake hands.))
71 Ten sälbär au adieu super
72 ***you too bye great***
73 ((The caretaker picks up his toolbox and leaves the flat.))

The tenant indicates twice that putting things back is his responsibility (lines 53 and 62). In both cases, he produces a verbal reaction to the caretaker's restitution of some objects in the kitchen, which he had to remove in order to make space for his repair work (45–46; 60). In the first case (53), he tells the caretaker that he does not have to do what he is doing, i.e., re-installing the mat and the chopping board. In the second case (62), he turns to the camera(person) and produces an ironical description of the caretaker, referred to with a third-person pronoun. Accompanied by brief laughter, this turn could be understood as mildly criticizing the caretaker for doing too much. While the first turn (53) was an invitation addressed to the caretaker, the second turn (61), accompanied by a look to the camera (Image 5), calls on the cameraperson to act as a witness to the caretaker's possibly excessive zeal (and stubbornness, since he did not comply with the first injunction (53)). It shows that caretaker and tenant may (more or less strongly) disagree on the division of labor, especially once the technical operation of the repair is complete (Fig. 4.5).

Fig. 4.5 The caretaker re-installs the chopping board next to the sink (© Video still: I. Strebel)

Another aspect of the division of labor appears in the final exchange of this sequence (49–59). As in his previous positive assessment (29–30), the tenant compares the current state of the tap with the pre-repair state, about which he says "it scared [me]". As mentioned before, the tap had been very loose, which made the tenant afraid of it breaking into pieces. The caretaker dismisses this fear (51, 56–57) by ironically arguing, "it cannot do more than break". After brief laughter, he adds "it is not that tragic". The caretaker indicates here that the tenant did not have to fear the tap being loose. The worst that could happen would be a break, which would not be tragic.

In this exchange, the tenant produces a positive assessment of the caretaker's repair by saying that his kitchen tap is no longer scary to him.[15] The caretaker can thus be credited with having restored a normal relationship with an essential artifact in the flat. The caretaker's response does more than that though. He is suggesting to the tenant that he too may be responsible with regard to the infrastructure of the flat. If a tap is going to break, then let it break (and call the caretaker), which appears as a normal course of events, compared with being frightened by your own kitchen tap. In other words, the caretaker gives more leeway to the tenant in his manipulation of the flat infrastructure. This final exchange before the closing sequence therefore shows the part of the repair work through which the figure of the responsible tenant and the competent and "liberal" caretaker, and their relationship to one another, are transformed. The caretaker does not only give instructions on the future use of the repaired tap, but also on how to deal with a potential broken tap in the future.

This final exchange of the closing of the caretaker's intervention addresses a central question for all caretaker work. Caretakers not only work in and for tenants' flats. What they do there is likely to modify the tenants' life in its most ordinary and routine aspects, just as a surgical act that takes a few minutes in the life of a surgeon may thoroughly transform a patient's life, for better or worse. Caretakers have to take this fact into account and accommodate the future use of the repaired thing. Once again, this is not just the "social dimension" of the caretaker work. It is constitutive of this work, because they do not repair inert broken things, but things in tenants' flats and lives. This might specify the professional ethics of caretakers. If we stick to the surgical metaphor, surgeons appear much less concerned with anything that falls beyond the strictly technical act. And patients regularly complain about surgeons' lack of concern for how to live with a repaired body.

This specific situation provides caretakers with a cognitive and practical authority on ordinary things. As we just saw, their intervention is likely to transform tenants' daily routines. We will now examine other examples of the way caretakers and tenants deal with the life-altering aspects of the situation.

Unrepaired Oven, Provisional Use

We will now return to a specific exchange within the previously examined unsuccessful attempt to repair the oven. We have seen how the caretaker made the failure accountable, both retrospectively, through the invocation of rust as a cause of the repair failure "*I can't remove the heating rod now*" and prospectively in making sure that he would come up with something, even if he did not yet know what: "*now I have to figure something else out*". In the following excerpt, the caretaker addresses another aspect of prospective accountability in explaining to the tenant (Ten2) and his wife (Ten1) how to provisionally live with a partly defective oven (Fig. 4.6).

Fig. 4.6 Leaving the flat with the oven unrepaired (© Video still: I. Strebel)

Transcript 5

Car guät also nöd guät aber
all right well not all right but
Ten1 ja ja
yes yes
Car aber iär chönnt ja immer no mit under-
but you can still with under-
Ten1 warten warten
wait wait
Car iär chönnt ja immer no mit under- und oberhitz
chönnt iär ja bachä
you can still with under- or over-heat
you can still cook
Ten1 ja guät
yes all right
Car also under- und oberhitz gaat
so under- and over-heat works
Ten2 ja ja
yes yes
Car isch nur umluft wo nöd gaat

	it's only the fan that doesn't work
Ten1	ja
	yes
Ten2	ich habe gesagt
	I said
Ten1	nur die luft XXX
	only the fan XXX
Car	ich ha nüt anders behauptät
	I never claimed anything else
Ten2	ja schon gut kein problem
	yes all right no problem
Car	((laughter))

The excerpt starts with a pun (1) that, by exploiting the double meaning of "all right", expresses the complex temporality that is addressed in repair closing. While the first occurrence of "all right" is clearly a structuration marker, the second one is part of a conversational self-correction that emphasizes that the intervention cannot be concluded with a positive assessment. This negative assessment is then tempered with the notification that the failure of the repair does not imply that the oven cannot be used at all. As the caretaker gradually makes clear (5, 9–10, 15, 19), what is broken and provisionally impossible to fix is the fan, but this does not prevent the upper and lower parts of the oven from working. While this is no news to the tenants, who claim to have already said as much (21, 23, 25), it seems to be treated as satisfactory (29).

This sequence shows that the caretaker is not strictly in charge of repairing what is broken but rather of restoring and maintaining the livability or viability of the flat to its inhabitants. While in this case, the problem for which the caretaker was called remains unresolved, the situation of the tenants with their problematic oven is being dealt with by the caretaker. At the end of this excerpt, the caretaker leaves the flat with the oven unrepaired, but with tenants who know that something is to be done shortly and that they can still cook in the meantime. The caretaker can thus be said to normalize things even when the repair fails.

Same Washing Machine, New Use

We have seen in the first part of the text that the failure of power supply to the washing machine was somehow surprisingly resolved by the caretaker, who immediately attempted to make clear to the tenant where the problem lay ("the key was not all the way in") and implicitly how it had been repaired. We will now see what follows on from that in terms of future use of the machine (Fig. 4.7).

Transcript 6

Ten	ah jetzt gots widär
	oh now it works again
((The caretaker moves back towards the key))	
Car	dr schlüssäl nit ganz din gsi
	the key was not all the way in
Ten	i hana zwoimal usagno
	I took it out twice
((The caretaker comes back to the tenant))	
Ten	[und widär innä tuä]
	[and put it in again][16]
Car	[ja wänn mä]
	[yes when you]
Car	also mängisch muäs ma no chli .
	sometimes you must add a little .
((The caretaker returns to the key))	
	druck druf gä
	pressure on it
((The caretaker gives the key two little pushes))	
Ten	ok in däm fall merci
	ok then thank you

After both parties notice that the machine works again, the caretaker returns to the key and formulates the cause of the problem (3–4). When he returns to the tenant, she replies with a brief description of how she had dealt with the key (6, 9). The second part of the description

Fig. 4.7 Formulating the cause of the problem (© Video still: I. Strebel)

overlaps with the caretaker's new turn (and move back to the key), which gives specific instructions on the manipulation of the key that is requested (11, 13, 15, 16, 18). The tenant ratifies this with thanks, which functions as a pre-closing of the whole intervention.

The tenant's reply (6, 9) to the caretaker's formulation of the problem can be understood not only as a refutation, but also as a form of justification: she describes what she did when she noticed the problem. She reports an action, putting the key in and taking it out twice, which apparently anybody, or in this normative environment, any responsible tenant unwilling to waste the caretaker's time, would have done before calling for help. The caretaker's interruption treats this argument as irrelevant: what matters is not how many times you put the key in and out, but how you insert it, in this case with two small pushes with the bottom of the palm of the hand, as the caretaker clearly and slowly demonstrates (18).

There is a form of upgrading of the accountability of the repair. The tenant's reply leads the caretaker to provide not a description of the cause of the problem (as in 4), but rather an embodied instruction as to how to solve it. In other words, what matters to the caretaker is not so much explaining to the tenant what caused the problem as instructing her on how to resolve it by herself in the future.

This repair is interesting not just because it is extremely contracted in time, with the repair even preceding the understanding of the problem. It is rather, but also partly as a consequence of it, that the repair is not a technical intervention inside the machine but the discovery of a specifically new use of it. What has been transformed is not the machine itself, but the way both the tenant and the caretaker will use it (and tell other people how to use it) from now on.

The cases we have examined so far may wrongly suggest that tenants are passive vis-à-vis the provisional or sustainable transformation of their privacy. But tenants do not, at least not always, just wait until their routine changes according to the contingencies of the repair work. They may also actively solicit the caretaker's authority in order to gain know-how that is relevant to their own use of the flat. We will briefly illustrate this with the following case with which we will conclude the second section of the chapter.

Repair as an Opportunity to Learn New Use

We now briefly return to the unsuccessful repair of the oven that we have already observed. At the beginning of the intervention, the caretaker had removed the oven door while talking to the tenant's wife. We have seen that rust had prevented the caretaker from unscrewing the oven's heating rod. Before telling the tenant that the repair has failed, he reassembles the oven. When he puts the door back in place, the tenant is sent by his wife into the kitchen to ask for a demonstration of how to remove the door. The caretaker accepts and starts a tutorial sequence (Fig. 4.8).

When the tutorial is complete, both the tenant and his wife offer justifications for their request: removing the door makes it easier to clean. The caretaker accepts the justification with a "no problem",[17] which contrasts with the repair failure that he then announces by stating another problem: "I don't remove the heating rod" (see analysis of Transcript 3).

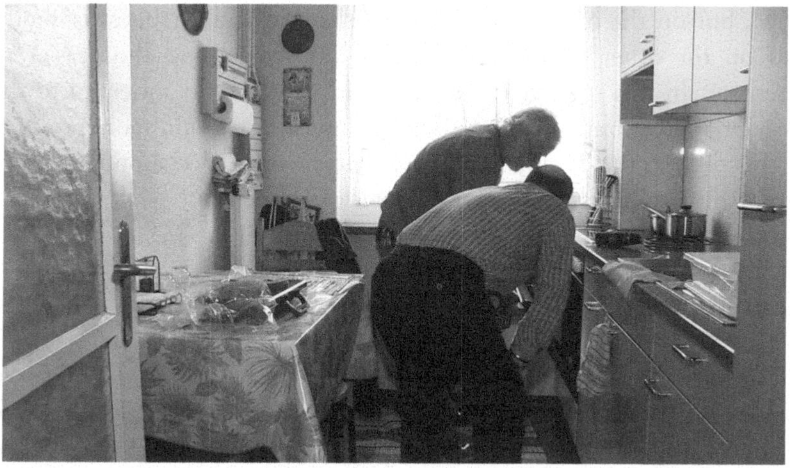

Fig. 4.8 Giving a tutorial (© Video still: I. Strebel)

This sequence shows that the presence of the caretaker can be exploited independently of the (in this case, unsuccessful) outcome of the repair. Caretakers have both practical (know-how, tools, etc.) and institutional (legal entitlement) resources that allow them to do things that tenants do not do, or do not dare to do, even in and to their own living space. The presence of a working caretaker can then be an opportunity to watch and even, as in this case, be taught a specific action. It is notable here that the tenants actively provide for the accountability of their request: it is for the sole purpose of better cleaning the oven that they take up some of the caretaker's time. In other words, they formulate a typical tenant's concern—cleaning the oven—as opposed to, e.g., replacing the heating rod, which tenants are neither expected nor probably entitled to do. At the same time, it displays the preoccupation of responsible tenants who share with the caretaker a larger concern for the maintenance of the flat infrastructure.

For such requests, the closing of the intervention appears to be the relevant time. When the caretaker puts things back as they were before, he is finished with, and thus will not be interrupted in, the technical phase of the repair. It may interrupt the phase of putting things back in place, but on the one hand, it is what that particular request is about, and on the other hand, this phase is less central and thus easier to interrupt. In this respect, the tenants can be said to be oriented to the sequential organization of the caretaker intervention, and more precisely in this case, to the closing as a phase that provides special opportunities for learning something about their flat's infrastructure.

Conclusion

Closure has been, and still is, an important concept in the sociology of technology. Laid out in an early text by Pinch and Bijker (1989), closure was conceptualized as involving "the stabilisation of artifacts" and the "'disappearance' of problems" (ibid.: 44). The word "disappearance" is emphasized because, as Pinch and Bijker write, "to close a technical controversy, one needs not to solve the problem in the proper sense of that word. The key point is whether the social groups see the problem as being solved" (ibid.: 44). Pinch and Bijker illustrate this point by highlighting two forms of closure: the first is "rhetorical closure" which we can consider as a true concept of the constructivist sociology of technology. Pinch and Bijker are convinced that the stabilization of technology can be achieved discursively, such as, for example, when consensus is reached in public discourse on the innocuousness of a potentially risky technology. Such closure is considered as rhetorical because it happens in the public sphere rather than on the testing ground.

The second form of how controversies come to a close is not by making the problem disappear, rhetorically or otherwise, but by redefining it. A classic historical example of "closure by redefinition of the problem" given by Pinch and Bijker is the inflatable and easily breakable bicycle tire, which was the technology—discovered in cycle races—that eventually triumphed, because the speed of the inflatable tire encouraged its material weakness to be overlooked.

Our chapter takes not the closure of controversies but the closure of technically instable situations through repair work as its starting point and confirms at least two elements of this early discussion in the sociology of technology. We have first demonstrated that ending a repair job in a tenant's flat has a very articulated "rhetorical dimension", which we (re)specified as being oriented toward the tenant and the livability and viability of the rented apartment. As in the early understanding of closure, a repair is closed when tenants are convinced or reassured that the job is done or will resume later in the case of failure. We can hardly imagine that, having just finished a job, a caretaker would leave the flat without saying anything to the tenant. We consider that a fundamental feature of repair work in residential buildings lies in this orientation toward the intersubjective, i.e., shared-in-common with the tenants, character of the achievement of the job, or of its eventual continuation. In relation to this, what we could also show is that the end of the technical job and the ways in which it is made accountable to the tenant are not to be separated into a technical and a social phase, but the technical and the social aspects of the closing work are organized in situ and they supplement and overlap one another in various ways.

The second parallel to this early discussion in the sociology of technology that can be drawn from our study of repair closing is the issue of "redefinition of the problem". We observed that it is in any case in interaction with the tenant that successful and unsuccessful repair takes a specific turn and becomes redefined. The orientation to the tenant is a potential translating device that contributes to further stabilize or destabilize the achieved successful or failed repair. As revealed in part one of the chapter, these features help to underline the STS understanding of closure as a clear-cut moment, which is productive in separating technology from a stable and an unstable situation. This is true if we look at the closure of repair work. In the first part of our analysis, we could observe and describe how a clear separation and asymmetry was established between the technical and the social components of repair work. The caretaker leaves the flat and its tenant with the now repaired and usable technology. The praxeological perspective is useful in that it helps achieve a better understanding of how technical artifacts become stabilized or destabilized through repair work. But while we use repair work to better understand how materiality is relocated in and through the process, what have we truly learned about repair as a practice? Regarding this question, part two of our study "Life after Repair" is particularly revealing as it shows, via a second look at the instances examined in part one, the ways that repair involves retrospective and prospective consideration of the human/material nexus.

We want to underline three characteristics of repair work that lead us to a different, less clear-cut, and more fluent understanding of the closure of the process. From this, we gain another appreciation of how repair relates to STS.

1. Rather than construing closing as a passage between the "before" and "after" phases of a technical situation, our praxeological analysis shows the way in which order is practically achieved from, and out of, a situation of "failure", "controversy", or "breakdown". In the cases of infrastructure and technology, this process can hardly be described in terms of "before" and "after" phases. While closure exerts a clear stabilizing effect, it also opens up a potential for rework. Repair does not sustain or complement infrastructure, but becomes an integral part of its quality. Should we not then dismiss the notion of closure? The answer to this question is no. However, we cannot limit ourselves to the comparison between two stable states and the interpretation of what one brings to the other. We rather suggest documenting how participants accomplish their everyday tasks and focusing on selected instances of change when social and material conditions take unusual and changing turns. Whether such a turn then is specified as achieving closure or another form of transition is up to the participants to decide.
2. Following from this, a fundamental relational aspect of caretaker work is that it does not only offer a new or restored object, but also a modified relationship to that object. This relationship not only includes instructions and advice on the future use of the object, but it also involves specifying how the object has changed. This is not done as a report on the technical intervention as in technicians' peer talk (Orr 1996) but as a specification of how the tenant's usual life in the flat will be altered by the repair process. A further feature that has been revealed is a form of "learning without teaching", where the tenant learns about the composition or working of the broken thing and how to use it. The tenant also learns that the reappearance of the same problem, or the appearance of another, will always be dealt with by the caretaker and is thus "not that tragic", as one tenant learned.

All these learning and instructional elements involved in closure suggest that it should not be understood as the end of something, but as a critical moment in which interactions become denser. Breakdown and crisis have long been used in STS thought as a heuristically useful feature for the analysis of the human/material nexus. Our observations invite additional examination of the way closure engenders and transforms not only objects, but also the relationships that future users will have with them.

3. Specifically, our cases of closure were all set in the private sphere of the residential flat, where caretakers work in the presence of, and in discontinuous interaction with, residents. The specific orientation that was highlighted in closing situations and the resources used to finalize and account for a job, such as the joke or the tutorial, respecify the tenant–infrastructure relationship of rented housing. In this respect, repair not only transforms social order but also incorporates it, in that if it were accomplished without orienting to the tenant, it would become another job. This feature makes caretakers' repair work truly distinct from other "technical" repair situations, such as, for example, when technicians repair or maintain an object or technology involving a colleague, an administrative agent or a specialist, such as an architect, facility manager, or engineer.

Acknowledgements We would like to thank the five caretakers for letting us shadow their work, as well as the tenants who kindly let the film crew in. We thank Sara Keel and Philippe Sormani for their insights into data sessions on this material. The research for this chapter was funded by the Swiss National Science Foundation (Grant No. 149478).

Notes

1. In our data, all caretakers are male.
2. Ignaz Strebel and Susanne Hofer (2014) Building Care: That's why our cities do not fall apart, online video (41 min. German/Swiss German, English subtitles), http://vimeo.com/ethwohnforum/building-care.

3. “Oriented” refers not only to spatial orientation, but also to an adjustment to any component of the situation. For example, a car driver's actions such as braking, stopping, or turning left can be said to be oriented to, among other things, other cars, pedestrians, traffic rules, weather conditions, etc.
4. This is the case even if the tenant is not present. One way or another, the caretaker will make sure that the tenant receives notice of the achievement of the repair.
5. The camera and sound operators stand at the kitchen door and thus constitute an obstacle to the tenant to enter the kitchen. It is nevertheless commonly the case that tenants do not stay with the caretaker for the whole duration of the repair and thus expect to be called when the caretaker is finished (or needs special assistance).
6. The peculiar participation framework, where the tenant, while not present, is both a ratified participant and within hearing distance, is oriented to by both parties here (Bovet and Strebel 2015). On the one hand, the caretaker does not expect a prompt answer, as indicated by the fact that he does not wait for it nor orients to its provisional absence as remarkable: as soon as he has formulated the summons, he resumes his cleaning business. The tenant takes some time to answer, but the fact that he answers before entering the kitchen suggests that he attempts to minimize the gap between the two pair parts. It can be noted that his answer deals with the short duration of the repair, which is an aspect of the caretaker's work that can be formulated before entering the kitchen.
7. This distinction is achieved, among other means, through changes in the participation framework (Bovet and Strebel 2015).
8. The standard oven in this housing, as in most similar housing in Switzerland, is an electrical oven with over- and under-heating as well as the fan-assisted function.
9. The tenant speaks a variety of broken German mixed with elements of Swiss German.
10. It could also exempt the tenants from responsibility in the breakdown of the oven but this was never at stake in this intervention.
11. Shortly after this excerpt, the caretaker offers apologies and assures the tenant that he tried his best.
12. See, e.g., Sennett (2013) and Ureta (2014).

13. The use of a kitchen tap is so routine and embodied a practice that a tenant can be commonsensically expected to be extremely sensitive to it.
14. That is the caretaker not only replaced the inner damaged part but also the worn-out joints.
15. It also retrospectively justifies the fact that he called the caretaker in the first place. Being afraid of his own kitchen tap appears a legitimate reason for calling a caretaker.
16. Square brackets delimit the portions of two turns at talk that overlap. Here it means that "und widär innä tuä" ("and put it in again") at line 9 is spoken in overlap with "ja wänn mä" ("yes when you"), line 11.
17. While the caretaker has "no problem" accepting the request, the video recording of the tutorial sequence (not transcribed here) shows him fulfilling it somewhat minimally: he removes the door, giving the tenant commentary, but does not give him the opportunity to try it himself.

Bibliography

Bearman, Peter. 2005. *Doormen*. Chicago: University of Chicago Press.

Bonnet, Lucie. 2014. Des préoccupations des habitants aux demandes du client: Gardiens et agents d'accueil en habitat social in *Le logement social en France*, ed. Charles-André Dubreuil and Patrick Néhémie. Clermont-Ferrand: Centre Michel de l'Hospital.

Bovet, Alain, and Ignaz Strebel. 2015. Claiming, maintaining and ascribing participation status in repair and maintenance work. A study of summons in concierge-tenant interaction. Paper Presented in Revisiting Participation Conference, 24–27 June 2015. University of Basel.

Brand, Stewart. 1994. *How Buildings Learn. What Happens After They're Built*. New York: Viking.

Edensor, Tim. 2011. Entangled agencies, material networks and repair in a building assemblage. The mutable stone of St Ann's church, Manchester. *Transactions of the Institute of British Geographers* 36 (2): 238–252.

Gamst, Frederick C. 1980. *The hoghead. An Industrial Ethnology of the Locomotive Engineer*. New York: Holt, Rineheart and Winston.

Garfinkel, Harold. 1967. *Studies in Ethnomethodology*. Englewood Cliffs: Prentice Hall.

——— (ed.). 1986. *Ethnomethodological Studies of Work*. London: Routledge & Kegan Paul.

Gold, Raymond L. 1952. Janitors versus tenants: A status-income dilemma. *American Journal of Sociology* 57 (5): 486–493.

———. 1964. In the basement—The apartment building janitor. In *The Human Shape of Work: Studies in the Sociology of Occupations*, ed. Peter L. Berger, 1–49. New York: Macmillan.

Gregson, Nicky, Alan Metcalfe, and Louise Crewe. 2009. Practices of object maintenance and repair: How consumers attend to consumer objects within the home. *Journal of Consumer Culture* 9 (2): 248–272.

Harper, Douglas. 1987. *Working Knowledge: Skill and Community in a Small Shop*. Berkeley: University of California Press.

Jacobs, Jane M. 2011. Urban geographies I: Still thinking cities relationally. *Progress in Human Geography* 36: 412–422.

Jacobs, Jane M., and Stephen Cairns. 2011. Ecologies of dwelling: Maintaining high-rise housing in Singapore. In *The New Companion to the City*, ed. Sophie Watson and Gary Bridge, 79–95. Oxford: Blackwell.

Jacobs, Jane M., Stephen Cairns, and Ignaz Strebel. 2012. Doing building work: Methods at the interface of geography and architecture. *Geographical Research* 50: 126–140.

Laé, Jean François. 2015. *Dans l'œil du gardien*. Paris: Seuil.

Luff, Paul, John Hindmarsh, and Christian Heath (eds.). 2000. *Workplace Studies: Recovering Work Practice and Informing System Design*. Cambridge: Cambridge University Press.

Marchal, Hervé. 2006. *Le petit monde des gardiens-concierges: un métier au cœur de la vie HLM*. Paris: L'Harmattan.

Omahna, Manfred. 2003. Der Hausmeister. In *Das ganz alltägliche Elend. Begegnungen im Schatten des Neoliberalismus*, ed. Elisabeth Katschnig-Fasch, 336–344. Wien: Löcker.

Orr, Julian E. 1996. *Talking About Machines: An Ethnography of a Modern Job*. Ithaca and London: Cornell University Press.

Payer, Peter. 1996. Hausmeister in Wien: Aufstieg und Niedergang einer Respektsperson. *Wiener Geschichtsblätter* 4: 1–19.

Pinch, Trevor J., and Wiebe E. Bijker. 1989. The social construction of facts and artefacts: Or how the sociology of science and the sociology of technology might benefit each other. In *The social Construction of Technological Systems: New Directions in the Sociology and History of Technology*, ed. Wiebe E. Bijker, Thomas P. Hughes, and Trevor. J. Pinch, 17–50. London and Cambridge: MIT Press.

Schegloff, Emanuel A. 1968. Sequencing in conversational openings. *American Anthropologist* 70 (6): 1075–1095.

Schegloff, Emanuel A., and Harvey Sacks. 1973. Opening up closings. *Semiotica* 8: 289–327.

Sennett, Richard. 2013. *Together: The Rituals, Pleasures and Politics of Cooperation*. New Haven and London: Yale University Press.

Stébé, Jean-Marc, and Pierre Sudant. 2002. *Les gardiens d'immeubles au coeur de la ville: figures, métamorphoses et représentations*. Brussels: De Boek.

Ughetto, Pascal. 2011. Gardien d'immeuble: sentir et ressentir. *Communications* 89 (2): 89–101.

Ureta, Sebastián. 2014. Normalizing Transantiago: On the challenges (and limits) of repairing infrastructures. *Social Studies of Science* 44: 368–392.

Part II

Networks

5

Mobile Phone Repair Knowledge in Downtown Kampala: Local and Trans-Local Circulations

Lara Houston

Introduction

The total number of mobile subscriptions in Uganda rose sharply, from 126,913 in 2000 to 16,356 million in 2012.[1] In 1999, Uganda became the first country on the continent where the number of mobile subscriptions surpassed fixed lines (Sheehan 2003: 60). The expanding mobile infrastructure has provided access to telecommunications to huge numbers of people for the first time, and their initial encounters with telephony have been mobile. Moments of adoption are often analytically privileged in the study of emergent infrastructures, as researchers describe the impacts of a technology and how it works for users. In contrast, this ethnographic study of mobile phone repair begins at the moment when handsets stop working. While international investment was funding the rollout of mobile masts across the country, repair workshops were also mushrooming. Repair technicians contribute to

L. Houston (✉)
Goldsmiths, London, UK
e-mail: lara@larahouston.co.uk

I. Strebel et al. (eds.), *Repair Work Ethnographies*,
https://doi.org/10.1007/978-981-13-2110-8_5

the adoption of telephony in Kampala, by undertaking the work of 'local improvisation' (Suchman 2002), but they also do much more. This chapter is part of a lively and growing research strand that makes deeper and more profound claims for the significance of repair as a crucial form of human interaction with technology (Graham and Thrift 2007; Jackson 2014). Repair is a way of reconnecting practices that fail through change, abruption and decay. Repair re-configures the reproduction of artefacts, systems and infrastructures: working to extend their durability, longevity and coherence. Repair technicians contribute to the appropriateness and sustainability of infrastructures, as they are made to perpetuate in particular social and material contexts.

Like other studies on repair, this chapter focuses on repair knowledge: the information, skills and experience that are developed by technicians in the process of learning repair, which extend and deepen throughout mature practice. My interest lies in the circulations of repair knowledge within local sites of repair, but more significantly, beyond them. I show how technicians' practice in downtown Kampala is shaped by connections to trans-local bodies of repair knowledge, largely accessed via the Internet. This interest was motivated by an independent technician called David, and takes his claim seriously that without access to the Internet there can be no repair. He explains:

> If they say no Internet, I also say no repairing, I get my bag and go home. I can't repair without Internet, that's out. Maybe if you are hardware repairing, but for me, new phones, new updates, new what? - Internet.
> David, interview excerpt, 19 September 2012

In order to understand David's assertion, I turn to the wider trans-local 'ecologies of knowledge' (Star 1995) that are known and accessible to technicians in Kampala: ranging from online interactions with the Internet via Google searching, towards more specific and sustained encounters with particular sites providing information about phone designs and strategies for repair. Two sites are particularly important within this chapter, firstly online libraries of firmware files, and secondly the virtual community of technicians known as GSM Forum.[2] Through the analysis of empirical cases, I show

how David and his peers in downtown Kampala search these sites for information about breakdown and repair, acting on relevant information, drawing it into local repairs, and extending their practice in the process.

As other scholars have highlighted, repair work is a highly contingent affair (Suchman 2007; Orr 1996); mobile handsets are complex artefacts and new configurations of breakdown emerge continually. Alongside commonplace faults, David grapples with oblique malfunctions, where phones demonstrate inconclusive symptoms that point towards multiple failures. Technicians also deal with the unpredictability of working with a diverse range of handsets from numerous manufacturers, spanning over twelve years of systems design (six iterations of Moore's law). David specialises in software repair work, which requires not only knowing multiple devices, but also locating firmware files specific to each handset model and version number. Like many of the research participants, David runs his own repair business, sometimes employing an assistant or apprentice. He connects to trans-local repair knowledge as a way to manage the variable character of repair work: to find ways through repairs that overwhelm his knowledge, and to locate highly specific firmware files.

Repair is made more difficult for technicians like David, because information about the design of mobile devices, strategies for repair, firmware files and tools to intervene into embedded software systems are routinely withheld by manufacturers, who choose to share this information with particular repair businesses through relationships of 'authorisation'. In the downtown area, I located four 'authorised' workshops. Although the relationships with mobile manufacturers and network providers were different in each case, these workshops generally had privileged access to schematics and repair documentation, and tools for accessing, testing and repairing particular sets of devices. This link between authorisation and access is replicated in the case of many commodity technologies, and describes a particular political and economic context for repair, where independent technicians work in the absence of specific knowledge about systems or artefacts and in competition with manufacturer-backed repair 'monopolies'. By virtue of their location in downtown Kampala,

David and his independent peers may be further marginalised by their peripherality to larger flows of technologies and knowledge, which are disproportionately centred in the global North. Additionally, they rely on an emergent telecoms infrastructure for Internet access, where getting and staying connected are difficult and frustrating processes that require non-trivial investments of time and money. In Kampala, accessing online sites is a comparatively new dimension to repair practice, because independent technicians have only had access to reasonably priced pay-as-you-go mobile Internet in recent years. In 2005, only 1.74% of individuals used the Internet in Uganda, rising quickly to 12.5% in 2010, the year my fieldwork began.[3]

In this chapter, I show how trans-local sites fulfil an important role in enabling the collaborative sharing and pooling of information about repair practices across geographically extensive technician communities. These ecologies of knowledge provide alternative sources for information about phones, knowledge about repair, and firmware files. How does connecting to trans-local bodies of repair knowledge help Kampalan technicians to manage the problematic nature of breakdown? How do trans-local sites such as GSM Forum help independent technicians to redress the asymmetry of manufacturer 'authorised' relations? The very limits of repair practice are at stake here—as David highlights, without these bodies of knowledge, phones would simply go un-repaired. From the workbench in downtown Kampala, I reflect back on the trans-local sites surfaced in the empirical cases, and what they might tell us about this aggregation of repair knowledge on a wider scale, particularly given that mobile phones are globally circulating commodities. Within the trans-local repair community, ways of solving particular forms of breakdown become known, stabilised and shared as 'solutions'. In a situation where over 1.8 billion phones were sold in 2014 alone (Gartner 2015) and over half the world's population has a mobile subscription (GSM Association 2015), what implications do circulating 'solutions' have for repair practices envisaged on a wider scale?

Literature Review

Repair knowledge has been of central interest for scholars of repair. One prominent example within the literature is Julian Orr's seminal study of photocopier repair work within the Xerox corporation, which is used as a key point of comparison throughout this chapter. Orr draws from studies of situated activity (notably the work of Suchman 2007) in elaborating on repair work. He explains that Suchman's 'claim for the fundamentally situated nature of activity is based on the premise that human mental activity is socially and materially located… This means, in part that actions, or practice must be understood with reference to the situation of their doing' (1996: 11). Breakdowns are often initially unknown to technicians and the situated practice of repair involves constructing an understanding of the problematic situation from within, and scoping possible ways to go forward. Orr characterises repair as 'a response to the fragility of available understandings of the problematic situations,' where 'understanding is fragile in that accurate information about the state of the machine is only sometimes available, and the meaning of available information cannot always be found' (1996: 2). There are two aspects of Orr's description of repair that are of particular interest here. Firstly, Orr highlights how the process of moving from unclear and unbounded problems towards a greater understanding involves reading the behaviour of malfunctioning devices, and searching for the meaning of these symptoms. Secondly, the fact that repair practices are situated means that these processes of searching unfold in relation to the situation at hand.

Suchman's work on situated action evokes the situation purely as a context for action. Without abstracting it into a 'container-like' form (Suchman 2007: 19), I want to emphasise how the setting for repair work varies across differently organised 'repair worlds' (Jackson et al. 2012). Many of the reasons that downtown Kampala is an interesting site for studying repair knowledge hinges on its difference from organisationally bounded sites within the literature (such as Orr 1996; Henke 1999; Sanne 2010), as I highlight in the introduction. In order to deal with the differences between these sites,

I draw lightly on Star's conception of 'ecologies of knowledge' (1995), partly to avoid loading the 'situation' with conceptual baggage that Suchman did not intend. At moments of breakdown technologies are unravelling; mobile phone parts are scattered across the workbench as technicians try to make sense of the trouble and refigure the device. At these moments of profound material entanglement, a priori categorical distinctions between humans and machines and associated pre-specifications of agency cannot accurately account for repair practice. Star's framing means 'refusing social/natural or social/technical dichotomies and inventing systematic and dialectical units of analysis' (1995: 2). The metaphor of an ecological space treats knowledge as deeply embedded in, and emergent from practices. Likewise, the representations of knowledge that I make in this chapter are anchored within practices of repair, and connected to particular repair worlds. Star also intends ecologies of knowledge to provide ways to account for absences, something that becomes important in characterising the lack of information that independent technicians work within. In doing so, this framing of knowledge also surfaces questions about power, access and justice that are touched upon in the conclusion of this chapter.

Finally, the ecological metaphor allows me to characterise knowledge in a way that I find consonant with the practices under study, incorporating a range of circulating repair resources such as schematics, firmware files, interactions on GSM Forum and artefacts such as the Nokia Hardware Library. This framing matters, because as Orr demonstrates, knowledge about repair is so much richer than information about machines. Repair cannot simply be understood as the result of the procedural application of knowledge, instead it is a process of continual improvisation between technicians, customers and machines (Orr 1996: 2). Even when technicians work in well-resourced organisational settings, they still come together to talk about troublesome breakdowns. One of Orr's central findings is the collaborative nature of repair work, and importance of narration and storytelling in the process of repair. Diagnostic work is performed through 'war stories' told by technicians about previous repairs that may be relevant to the current problem under discussion. Orr explains that 'war stories told during difficult diagnoses are doubly situated, first in the context of their origin and then in that of the telling and possible application, and the comparison of the two situations is the point of the telling' (1996: 127). In addition to offering possible new understandings of the difficult case, these descriptions also serve to circulate repair knowledge within this particular occupational community and in doing so perform identities of mastery, a contribution to the social ordering of this repair world. I return to these themes later in the chapter, when discussing interactions between technicians on GSM Forum.

Two recent additions to the repair literature are also worth highlighting for their contributions to the characterisation of trans-local connections within repair practice. The first, by Ahmed, Jackson and Rifat (2015), describes communities of mobile phone repair in Dhaka. The authors trace how repair knowledge is disseminated locally through apprenticeship relations, as novice repairers are trained by mature practitioners. Trans-local knowledge is introduced by technicians who have taken repair training abroad, who bring their skills and expertise back to Dhaka, where it is passed on to apprentices. Ahmed, Jackson and Rifat also highlight how repair practices are shaped more generally by trans-local flows of information, technologies and tools, and they describe how technicians in Bangladesh make efforts to situate and adapt these, while also nurturing a local repair ecosystem. This chapter builds on their work by engaging in a sustained treatment of how trans-local knowledge comes to be accessed via the Internet, and embedded locally. The second study by Pollock et al. (2009) describes 'post-local' forms of repair in systems of virtualised technical support. In the troubleshooting and repair of large-scale software systems, customers and support staff are increasingly located in geographically distant places, with communication taking place through a web portal. This set of repair practices complicates questions of locality more profoundly, as it can no longer be mapped onto a single geographical place, but rather takes place across an 'extended' situation, in an evocative example of how repair knowledge is generated across networks, between sites.

'Authorised' Sites and Knowledge

Sloping gently down from the relative order of Kampala Road is the bustling locale known as 'downtown'. It is one commercial heart of Kampala, and trade is visible everywhere. The main streets are largely populated by a medley of shopping arcades four or five stories high, divided into small premises, and ringed with walkways open onto the streets. Outside, vendors chat in the sunshine, as people stream uphill towards the centre of the city. Amongst all this activity, the infrastructure of mobile telephony can be traced. Women selling mobile phone pre-pay credits or 'airtime' congregate under large, shady umbrellas that punctuate the street. They are a public telephone system, offering calls on mobile payphones to passers-by, and keeping those with their own mobiles topped up with credit. Looking upwards from street level, the buildings are adorned with banner advertising, emblazoned with the bright colours and catchy logos of mobile manufacturers and network providers. Inside the shopping arcades, clusters of mobile-related businesses sell, personalise and repair mobile phones. At each turn, rows of mobile handsets new and old stand to attention in illuminated display cases. Repair technicians are visible at their workbenches, surrounded by unhoused phones and repair tools.

This study draws on six months of ethnographic fieldwork undertaken over three years, between 2010 and 2012, in fourteen repair workshops within the downtown area. Across workshops, there were divides in practice: five specialised in hardware repair, three specialised in software repair and seven more performed both types. As David notes, online resources were particularly important for software repair practices. Eleven workshops in total were connected to the Internet: all of the businesses that engaged in software repair, plus one hardware workshop. A majority of the repair businesses in downtown Kampala were small, independent enterprises, owned and run by a single technician, sometimes employing an assistant or apprentice. These workshops tended to be located in shared shops within the shopping arcades of downtown, which had been subdivided down into desk-sized spaces. As a counterpoint I also looked for workshops that diverged from this norm, and found four workshops that were different in size, ownership and authorised affiliation. All four maintained relations of 'authorisation'; two with mobile manufacturers and two with network providers. Three of these workshops were part of much larger businesses; the service centre of an East Asian brand was one of over seventy others dotted across the African continent; the workshop authorised by two multinational brands was part of a wider mobile phone retail chain which had outlets across East Africa. The third enterprise was the only one of the four that was owned and run by a technician, and it had scaled up by installing and managing concessions in the retail outlets of a network provider, in addition to maintaining a standalone repair workshop. The fourth and final workshop was the only small-scale 'authorised' business, employing one technician and one salesperson. It was owned by a Kampalan businessman and repaired warranted phones for a particular network provider.

Relationships of authorisation are important conduits for the travel of trans-global repair knowledge from centres of mobile design, to approved and sanctioned locations of repair. However, the term describes four very different sets of connections, with their own limitations and dynamics of control, which I briefly outline here. Often, these linkages were developed in order to fulfil warranties, the guarantees given to consumers by mobile manufacturers and network providers that any faults would be repaired (or the handset replaced) within the first twelve months of purchase. In the service centre run by the subsidiary of a fast-growing East Asian manufacturer, repair knowledge circulated initially via migration, as two technicians had travelled in person from the mobile phone factory to develop the repair facility based in Kampala. This marks an intimate relation between sites of manufacturing and repair. Technicians within this workshop drew on a rich ecology of repair knowledge, including tacit and embodied aspects of repair demonstrated by the founders, in addition to schematics, tools and firmware files which were updated via email and Internet file transfers. The second enterprise was authorised to retail and repair the phones belonging to two multinational brands, which provided technicians with access to online knowledge portals, where they could access information about handsets and repair tools. These also provided mechanisms for repair training, such as online courses (which technicians were required to take) and a question and answer facility for troublesome repairs. However, within this workshop, technicians also repaired phones made by other manufacturers, leading to an interesting juxtaposition of 'authorised' and 'unauthorised' practices.

In contrast, the two workshops 'authorised' by network providers had limited access to repair knowledge. The small workshop undertaking warranty repairs was provided with software tools and spare phones to cannibalise for parts, but no ongoing support. The technician-owned business had no special access to information or tools at all. However, the owner paid for his staff to take online training courses from a major manufacturer, and travelled with several technicians to a lab in China to have the quality of their repairs certified. In an inversion of the other sites, connections to trans-local knowledge had to be demonstrated as a condition for getting the contract for concessions, and for the repair business to become 'authorised'.

Outside of these authorised linkages with particular workshops, information about the design of devices, tools to intervene in them and strategies of repair are withheld by manufacturers. In the process of trying to develop a completely open source mobile phone called FreeRunner, software programmer Harald Welte explains:

> Billions of cell phones are being used every day by an almost equally large number of users… Despite being an openly published international standard, the architecture of GSM networks and its associated protocols are only known to a relatively small group of R&D [research and development] engineers. Even less public information exists about the hardware architecture of the actual mobile phones themselves…. (2010: 1)

The 'openness' of the GSM standard does not mean that information about the design of GSM devices is freely available outside of what Welte describes as a 'select club of cell phone makers' (2010: 10). In his comments, Welte describes how an absence of information restricts and contours attempts to intervene in the mobile market, in his case to produce a radically 'open' handset. However, this account also evokes the constraints placed on the work of repair.

Relationships of 'authorisation' facilitate the trans-local circulation of repair knowledge from centres of design to particular accredited locations of repair. Having privileged access to these knowledge flows is part of the value claim implicit in the 'authorisation' of workshops, in performing repairs that are visible, and legible to the corporation. These dynamics show how repair work is intimately bound up with design practices, yet sits in ambivalent relation to them. In contemporary cycles of technological obsolescence, manufacturers offer limited ways to engage with repair, structuring the circulation of 'official' knowledge in a way that produces and maintains asymmetrical ecologies of knowledge, and by extension deeply uneven power relations. Independent technicians work within an absence of 'official' knowledge about these ubiquitous devices. These proprietary closures raise many challenges for independent repair work. Finding information about breakdown and repair is a large part of the work of dealing with difficult cases. When information about the design of devices, and their modes of breakdown is not available within the organisational structuring of repair, searching for it becomes a problematic dimension of repair work. Where independent technicians are competing with 'authorised' workshops, it becomes an added burden as independent technicians are searching longer and harder for information than their authorised counterparts, as the case elaborated in the next section demonstrates.

Online Resources in Diagnosis: 'Nokia 6500 Restart & Off'

This case from my empirical fieldwork describes the work of searching for firmware files in the context of a problematic repair. It takes place in the basement workshop of an independent technician called Jason, who works alongside his younger half brother and apprentice Isaac. Every other trader on this floor sells car tyres, but despite the dim light, Jason has made a comfortable home in his concrete niche, with illuminated counters displaying spare parts for sale, and his favourite country music playing through computer speakers. It was a home where I was welcome to perch on a spare stool, to watch and to listen to the daily rhythms of repair. No facet of everyday life was exiled from repair shops. Friends and family visited, wives and children came to call. News and views were exchanged over breakfast and lunch. It would have been hard to ignore this '*mzungu*' (Kiswahili for 'foreigner') so different in nationality, race and gender, seated side-by-side with technicians behind the workbench. My residence was a mix of quiet watching, and lively performance. Stories that animated the differences between 'here' and 'there' would not only elicit hoots of surprise and laughter from technicians, but also gave rise to deeply tense and uncomfortable moments, as very different life opportunities emerged and were negotiated in conversation, particularly cohering around my mobility. The accounts within this chapter emerge from the richness, yet partiality of this position (Haraway 1988).

Jason had been in the mobile repair business since 2001, and he was the most experienced repair technician that I observed at work. Although new to mobiles, he was not new to repair, having previously worked as a watch fixer. This was not the most common route into repair amongst the technicians I spent time with. A more frequent pattern was for young people—predominantly men—to take up mobile repair during their studies at high school, college or university, or in the precarious years following graduation. A career as a repair technician had rarely been the explicit goal of the participants in my research. As they started to make their way in the world as independent adults, they were often inducted into repair through kinship ties between brothers, cousins and close male friends. In a bleak job market, starting a business as a technician was a way to begin living: to fund more education, to stay afloat in the city, or to support a young family.

Because of his broad skill set and extensive experience, other technicians solicited help from Jason on a daily basis. Despite running their own businesses, technicians regularly worked together, particularly within the cluster of repair-related enterprises in the downtown area. In this way, knowledge of repair circulated locally, within co-present relationships between technicians who consider each other friends and colleagues. Unlike larger technologies such as photocopiers, mobile phones are small enough to fit into pockets and be carried across the congested landscape of downtown. Technicians exchanged not only stories about breakdown, but also the broken phones themselves, enabling them to work together in ways that ranged from the collaborative to the transactional. Friends and colleagues might freely offer each other help when a particular repair failed, or lend each other cables, parts or tools. At other times, technicians would pay each other to undertake repairs, charging low rates for colleagues so that each party would profit from the repair. These transactional relationships often served to bridge divides in practice, as less experienced technicians sought out those with more highly developed skills. Technicians who only performed hardware repairs often maintained networks of software repair technicians that they sub-contracted when a software repair came into the workshop (and vice versa). Local repair knowledge enfolded value judgements about which technicians were good at particular repair techniques, and calculations about which colleagues might charge least for help. Authorised workshops also subcontracted work out to independent technicians, but there was no reciprocation, meaning that authorised knowledge does not come to circulate across local networks unless the technicians employed within these workshops change or leave.

Introducing Software Repair

This case begins with a software repair, which is usually solved by a process that technicians call 'flashing'. The motherboard of the mobile phone contains a flash chip which holds the firmware that allows it to run. Firmware files are code objects that perform a range of functions, from guiding the basic hardware functions of the phone, to booting the operating system which the phone user interacts with. They are produced by the manufacturers of the handset, or in some cases, the producer of the main system processor chip. Technicians use tools to access the embedded systems on board a phone, erase the existing firmware files on the flash chip, and then re-write new copies (Fig. 5.1).

This photograph (taken in Jason's workshop) shows one assemblage of software repair tools that Jason has brought together to access the embedded systems of a malfunctioning phone. The phone is in his

Fig. 5.1 A software repair tool assemblage, Jason's workshop (© Photo: L. Houston)

left hand, connected to two power cables which allow it to boot without a battery. The third cable carries data, linking the phone to the tape-covered hardware device in the centre of the photograph. This peripheral is known to technicians as a 'flasher box' or simply a 'flasher'. The flasher contains a programmed circuit, which establishes a connection between the phone and the computer, acting as a hardware interface. On the computer screen, a window is visible. A software program operates with the flasher and across the assemblage to enable a range of interventions into the mobile phone. Through this interface, technicians can use self-test procedures, reset factory settings, re-build serial numbers, and remove SIM locks, amongst other functions. On-screen, a log window shows a stream of scrolling text, which reports back to the technician the progress of the ongoing action, and the status of various components within the phone. The connections between the computer, software, flasher box and phone must be active before any interventions can begin. Most 'authorised' technicians have access to manufacturer-made tools to undertake similar tests and repairs. However, there is a small, but lively and competitive global market for tools produced by third-party developers that offer independent technicians access to similar functions. Each flasher box and software system addresses particular brands and groupings of phones, requiring technicians to purchase and use multiple systems. As an independent technician, Jason uses third-party tools that he first researches online, and then purchases from his friend Ahmed, an importer.

Software repair tools allow technicians to read the firmware files currently installed on a phone, and to make backup copies. But if the phone is malfunctioning, re-installing the same firmware file is likely to recreate the same problems. Instead, technicians have to find a stable, working copy of the file from other sources. Hopefully, writing these onto the phone will eliminate any malfunction. Firmware files are regularly iterated by manufacturers, so technicians must carefully negotiate the multiple versions that may be appropriate to a particular phone, as writing the wrong file can cause the handset to fail entirely.

Finding Firmware

Problems locating appropriate firmware have prompted Ibra to walk to Jason's workshop just a few streets away, in search of help, as I describe in my field notes below:

> Jason's close friend Ibra came into the workshop with a Nokia 6500 phone. He wanted to flash the phone, but explained that he didn't have the firmware files. He demonstrated the phone's problem - on start-up it just showed the Nokia logo, and then went black. Jason connected the phone by USB to the Turbo Flasher box, and using options on the software window, he tried to flash the phone using the firmware files that he had to hand on his computer. The rolling software log on-screen notified Jason that this process had failed. He connected the phone to the hardware box using a different connector - FBUS - and tried again, but that also failed. Then, Jason used Windows to search his own computer for matching files, using the internal model number of the phone "rm-240". He was not satisfied with the search results, and so he went online to the Shrak Mobile firmware file store, and browsed through the sections seeking the latest uploads for the rm-240 phone. Next, he searched on Google for "rm-240 latest flash file" and afterwards "rm-240 v10.60" and then "rm-240 v12.35"... Field notes, September 7, 2012

The first trans-local knowledge encountered here is circulating firmware files; code objects that are worked directly into repair practices, enabling them to take place. Part of Jason's expertise as a technician is keeping abreast of current sites where these files are hosted. It is likely that the firmware files are limited by copyright protection and End User Licensing Agreements (EULAs). However, the legality of the sharing of these artefacts is an open question within the technician community. Jason visits the Shrak Mobile website first, which is a large and well-organised repository of firmware files, schematics and mobile phone manuals (offline at the present time). Although files are freely available on this site, the 'cost' of the download is expressed through the affective texture of experience of the service. Files are available for download for free using the 'regular' service, but in using these technicians face a raft of restrictions: delays before downloads, delays between downloads, slower downloads, no 'resume' function for failed downloads, and both banner and pop-up advertising. A 'gold' subscription, paid for by credit card removes these obstacles. Downloading firmware files is an elaborate and time-consuming form of work which shapes and organises the workflow within the repair shop, a dynamic that these sites exacerbate. In this case, Jason abandons Shrak in favour of Google. He performs three searches, moving from an identifying question about the latest files, to searches seeking out specific file variants, and scans the results of each search, which he finds unsatisfactory. He then moves on to a website called GSM Forum:

> where he searched for 'rm-240 flash files' and then 'rm-240 v10.30', which generated a list of threads. He found a thread within the results called 'Nokia 6500 restart & off'. He opened it and started reading. It was about a hardware problem. Further on in the thread there was a picture and a brief instruction to remove the hands-free IC (Integrated Chip)... Field notes, 7 September

GSM Forum is an online forum and virtual community for repair technicians. Message boards are one of the oldest and most recognisable formats for communication online, with a genesis in the bulletin boards of the early 1970s Internet age. This forum follows the standard hierarchical format, with a front page that lists multiple sub-forums, each with a different topic. Members interact asynchronously by posting written messages on 'threads'. Guests can freely view the site, but member registration is required in order to create a thread or to post a reply. At the time of writing, the front page of the site lists 2,386,078 members from a wide variety of different countries, although the only language permitted is English. Large areas of the forum are dedicated to support, including two separate sections dedicated to hardware and software repair, which are laid out in a taxonomy of phone brands. Within these areas, technicians can post details of problems they are encountering with the repair of a particular device, which are open to responses from other members, and are then indexed for future searches. However, searching was only one way that technicians interacted with the forum; as a technician called Stephen explained, 'every day I have to come early, on the forum, login, discuss, we see what's new, and you keep on like that' (Interview excerpt, 21 September, 2012). Technicians consulted the News section to keep track of updates about GSM technology, to follow discussions about particular tools, and to look out for suggestions that were relevant to their repair practice. Technicians considered themselves members of the GSM Forum community.

Searching GSM Forum

After fieldwork I reconstructed the searches of GSM Forum made by Jason. Querying for rm-240 flash files brings up a huge range of threads; a total number of 477 at the time of writing. After scanning the results from this search, Jason again moves to search for a specific software file, this time v10.30. In my reconstruction this yields a much smaller set of results—only 23 threads in total. Towards the bottom of the results is a thread called 'Nokia 6500 restart & off'. This thread title describes the behaviour of the phone Jason is working on, which partially starts up and then switches off. Of all the search results on Google and GSM Forum that he has scanned, this is the thread that Jason

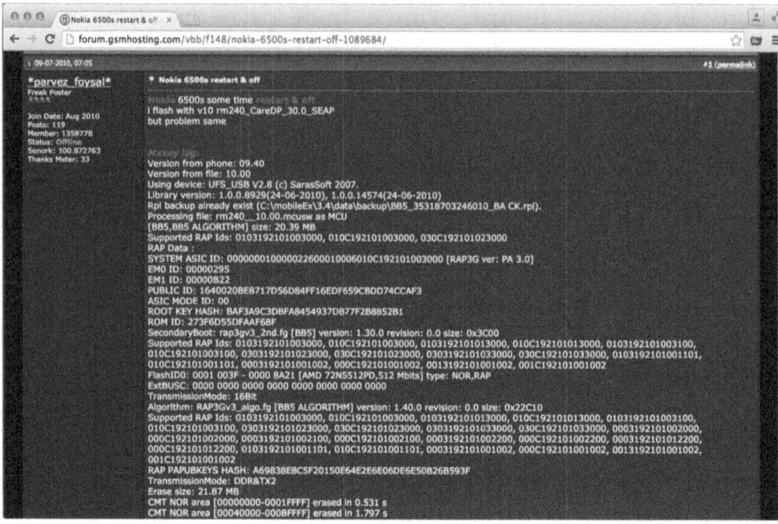

Fig. 5.2 An image posted on the GSM Forum thread 'Nokia 2600 restart & off', by the forum member *parvez__foysal* (© Photo: L. Houston)

chooses to open and read. It was started in September 2010, and spans a single page (Fig. 5.2).[4]

The first part of the post reads:

> Nokia 6500s some time restart & off
> i flash with v10 rm240_CareDP_30.0_SEAP
> but problem same
> GSM Forum, September 7, 2010

The original poster, *parvez_foysal* describes the model of the phone, the problem as he sees it, and the action he has already taken. This happens succinctly within the first three lines, characteristic of communication on the forum. He includes specifics about the flashing process, such as the filename of the firmware version that he used. Secondly he has posted the log file generated by the software during the flashing process (visible on the image above but not reproduced here for brevity). This is a pragmatic concern as it allows other posters to

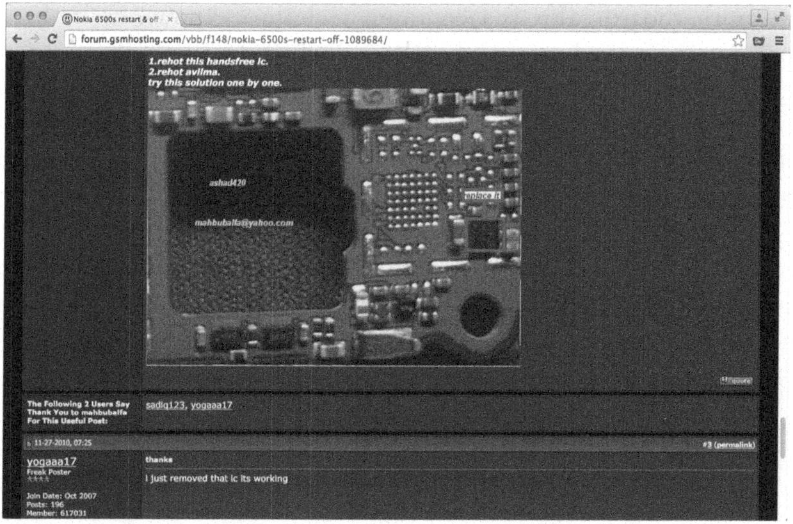

Fig. 5.3 An image posted on the GSM Forum thread 'Nokia 2600 restart & off', by the forum member mahbubalfa (© Photo: L. Houston)

verify the action he has taken, but it also conforms to what constitutes a 'good' posting on this forum, where members are quick to point out breaches in this community norm. Beneath *parvez_foysal*'s post is a reply from another member, named mahbubalfa. He quotes the original post, and beneath writes the following instructions, and then posts an image (Fig. 5.3).

The text reads:

1. rehot this handsfree ic
2. rehot avilma

try this solution one by one.

Here mahbubalfa sketches out a program of action for the original poster *parvez_foysal* to undertake. Firstly, to use a soldering heat gun (known as a 'blower' for the way it blows out a stream of hot air) to heat the Integrated Chip (IC) responsible for the hands-free function of the phone, to ensure that the chip is fully connected to the motherboard. This is a common technique known as 're-heating'. Secondly, to heat another IC called Avilma, which works to regulate power. All of the instructions given by mahbubalfa are brief, and they fold in technical language and terminology specific to mobile phone repair. The third action is shown in the picture, where an IC is marked up with the instruction 'replace it'. Re-heating an IC is a simple action while in Kampala, replacing an IC is considered an extremely skilful intervention. The very visuality of the pictorial instruction provides a more detailed quality of information, and a different orientation towards the suggested action.

> Jason looked at the post for a long time and decided to try it. He heated the board with the blower and picked off a tiny IC with his tweezers. After the phone had cooled, he restarted it, and it booted up perfectly. He put the phone back together and slipped the removed IC into the case. He said Ibra would replace it with another IC later, and set the phone aside. I thought the repair was great, and said to Jason that it would have been so difficult to read this problem from the phone's symptoms. He replied: "that's how important this forum is"… When I got back to the workshop Ibra had come for his phone, and Jason had also given him on a USB drive the updated flash files for the 6500, so that next time he could flash this model. Field notes, 7 September 2012

Although mahbubalfa's instructions are staged, Jason moves to the action that he feels is most important.

This repair demonstrates how diagnostic and repair work are inseparable (Dant 2010), as the diagnosis is stabilised only after the first intervention, which also effects a partial repair. Jason's collaborative role ends when he is sure he has identified the cause of the problem, and he leaves the final act of replacing the IC to Ibra. He also passes on the firmware files that Ibra originally came for, to add to Ibra's archive, and enable him to undertake future repairs of this model (an example of local circulations).

The Work of Searching

Problems with devices are often initially unclear and unbounded. In this case, the symptoms presented by the Nokia 6500 phone (which on start-up briefly showed the Nokia logo, and then went black) were very general, and could have reflected many kinds of trouble. The situated practice of repair involves moving towards a clearer articulation and bounding of the breakdown. Reading the phone's malfunctions, Ibra made a provisional diagnosis of software trouble, and after finding that he did not have the corresponding firmware files in his archive, the problem then became locating these. Ibra's deep and reciprocal friendship with Jason provided one possible way to move forward with the repair, by opening up access to Jason's archive of files. When Jason failed to flash the phone using the files in his archive, Internet browsing and searching opened up another possible way to locate newer and hopefully more appropriate files. Repair involves searching, which happens across the emerging situation, involving both local, co-present and trans-local online knowledge. Jason's understanding of the repair changed within this process, when he encountered the thread by *parvez_foysal*, generated by searching for the firmware file rm-24 v10.30. The original poster *parvez_foysal* had already performed the action that Jason wanted to undertake, and upon reading the thread, Jason recognises that mahbubalfa's suggestion is also applicable to his repair. Searching GSM Forum added information to the situation that allowed the ambiguous malfunctioning of the phone to be accurately interpreted.

In technicians' accounts, the term 'searching' glosses a series of activities that technicians undertake to achieve local repairs. Technicians talked about the GSM Forum website, and the work of searching in particularly interesting ways. A software repair technician called Stephen described the information found on GSM Forum as 'solutions,' a term that was commonly used by technicians in downtown Kampala, and also in trans-local knowledge artefacts themselves, such as the Nokia Hardware Library described later in the chapter. He explained that:

> …anyone can log in, search for any criteria they want. All solutions we keep them there, yes. Anything from the old phones from when we started, solutions we keep them. Search and get solutions, there and then, and then you are through with the problem. Interview excerpt, Stephen, 24 September 2012

For technicians in Kampala, 'solutions' represent possible trajectories of action. They are solutions to known, but loosely described problems that are not certain to produce repairs in an unclear repair situation. A few days after the Nokia 6500 repair, I asked Jason more about the importance of GSM Forum to his repair practice. He explained:

> I just go there and search the forum, they give solutions, I try them out, give solutions, try them out. So, what I have got from those people is very good, and I appreciate them. Interview excerpt, Jason, 11 September 2012

Jason's account is much more evocative of the process of trial and error involved in locating a relevant 'solution' and using it to construct a repair. Technicians query GSM Forum with search terms relevant to a phone, a symptom, or a hypothesis, and scan the search results. Online searching involves generating 'solutions', assessing their relevance, and deciding what (if anything) to take forward. Searching might locate firmware files that enable particular repair actions, or add new information to a repair situation, changing the conceptual understanding of a breakdown. However, it also involves acting: trying out part of a solution and then moving on to another one. The term searching obscures the work of alignment and construction that actually occurs when trans-local repair knowledge is drawn into local instances of breakdown and repair. Searching in situated practice complicates conventional divisions between understanding and acting that are inseparable in repair.

The 'solutions' that technicians describe tell us something interesting about the forms of information that circulate on GSM Forum. 'Solutions' are kernels of repair practice that are motivated towards resolving a particular issue. Like Orr's 'war stories,' replies on GSM Forum aim to offer new, contextually relevant information into the repair situation. Solutions are exchanged between posters and commenters in interaction, oriented towards a shared technical practice. However, unlike Orr's work, these 'stories' are extremely truncated. Community norms ensure that 'solutions' are directed clearly and concisely towards action, so that they can easily be searched and shared. 'Solutions' do not fulfil the same purposes as 'war stories' with regards to making claims to mastery within the community, but arguably the social role of repair work is negotiated across the GSM Forum as a whole. It is also interesting to highlight what travels within a 'solution'. The instructions given by mahbubalfa are not comprehensive step-by-step descriptions of how to achieve the repair, rather they enfold three different strategies for repair ordered by increasing intervention. They involve textual and visual notations, which require knowledge of repair techniques to understand and richly embodied skills to action. 'Solutions' do not get at the bodily knowledge of sight, sound, touch and gesture that also constitute material interaction.

'Solutions' act in two ways: they are first and foremost solutions in response to an original problem; in the case of the Nokia 6500, mahbubalfa's post was itself a response to *parvez_foysal*'s request for help. Yet these threads have impact far beyond the original question and answer. Threads in the support section of GSM Forum describe troubles long remedied or abandoned, yet because they are open to searching, these still retain value as interactions, and remain open to the possibility of being built upon further. In the virtual community of GSM Forum, 'what is important… is the communication itself - the shared informational artifact that is created by the participants' (Erickson 1997: 3). Interactions between technicians contain directed information about repair, 'solutions' that can be worked into similar situations. Within the support sections of GSM Forum, interactions between technicians accumulate through time into a huge body of shared knowledge. Made searchable, this is a huge and constantly growing, continually shifting archive which is an incredibly powerful resource for other technicians. Arguably, solutions only become stabilised and known as solutions over time within the community through the reiteration and refinement that the search function enables. The thread 'Nokia 6500 restart & off' has one further post, later in 2010, by member called yogaaa17, who replies to say the 'solution' has worked for his phone. Using the 'thanks' button another member has 'thanked' mahbubalfa's post; through these ongoing interactions, the veracity or usefulness of the post takes on a different weight.

Globalised Phones, Trans-Local Repair Knowledge

This case starts to open up questions about the geographical extensiveness of repair practices, within and beyond the GSM Forum technician community, in the context of mobile phones as globally proliferating commodity technologies. As I have emphasised, technicians contend with all of the variation that this brings to the practices of repair: different designs across manufacturers and models of handsets, multiple versions of firmware. In response, technicians cooperate and share; they interact with each other locally within downtown Kampala, but also at a distance on GSM Forum where their interactions aggregate into a shared archive of repair knowledge. This is a dynamic in which distributed technicians centralise information and knowledge to manage the contingencies of repair. However, technicians' sharing of knowledge depends on the standardisation of phones. Jason was able to take up mahbubalfa's solution, because the class of objects labelled Nokia 6500 conform to the same design. This is not a suggestion that all Nokia 6500 phones behave in the same way. As Orr observes, the fact that machines behave idiosyncratically is one reason why repair is problematic (1996: 2). However, the suggestion to remove a particular IC will be actionable in any location where skilled technicians have the necessary tools. This is an interesting dynamic of repair, because just at the moment that phones are coming apart in breakdown, technicians appeal to the stability and fixity of their designs.

At the same time, this shared knowledge of repair presents a very different way of knowing phones than information that is generated at moments of design. Through the sharing of problems on GSM Forum, idiosyncratic forms of breakdown within particular devices can be seen in aggregate, giving a wider view on patternings of breakdown and their corresponding solutions. There is a shared sense of how phones are likely to break, their possible weaknesses, and the durability of materials or connections. This wider framing can be more clearly seen in instances where repair 'solutions' are brought together, such as the Nokia Hardware Library (NHL) by BlackAttack, which I encountered within many of the repair workshops in downtown Kampala. This interactive Flash software program aggregates a range of pictorial solutions like the one posted by mahbubalfa. As an artefact, it has been formatted to travel widely, and it has been uploaded and shared by a wide range of Internet users on GSM Forum and beyond. The image below shows the front page of the program, with icons at the top that can be used to contact BlackAttack, find out more about the models supported, and turn off the automatic loop of muzak (Fig. 5.4).

A phone model can be selected from the drop down menu visible on the top right of the image. Then a second drop down menu appears, from which a problem area can be selected. An example of a solution for an N91 display problem is below, labelled simply 'display'. Unlike

Fig. 5.4 A screenshot of the front page of the NHL (© Photo: L. Houston)

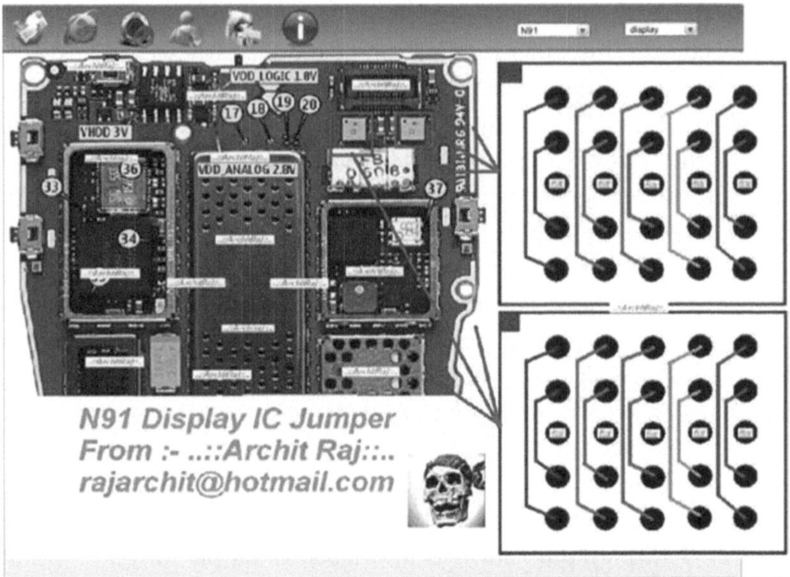

Fig. 5.5 A screenshot of the NHL N91 display solution screenshot (© Photo: L. Houston)

threads on GSM Forum, there is no contextual information about forms of breakdown and repair action already taken. Two ICs are highlighted on the motherboard image, leading to two boxes showing how copper wires must be laid in order to keep the phone working, when these parts are removed (Fig. 5.5).

These images have been produced by many different technicians, and are compiled rather than authored by BlackAttack. The front page of the NHL contains a request to upload any missing or updated solutions to BlackAttack's server, which is accessible through the menu bar. When I asked Jason if he knew anything about the makers of the NHL, he explained:

> ...many of them are on GSM, but it's not made by just one person, no. They just gather solutions, so and so got this solution, so and so got this solution and some people test them and see that they work. Interview excerpt, Jason, 11 September 2009

Images tend to include the name or moniker, email address, and sometimes a photograph or icon of the producer, raising questions about authorship. There is no sense that the person named on the schematic is the author of the 'solution', but they are the creator of this particular iteration of it, and seek to be strongly identified with it. Perhaps in this sense, 'solutions' connect to identities of mastery, although in a drastically different way to the storytelling observed in Orr's 'war stories'. Often it is not possible to discern much additional information about the authors, but a few of the images reference the locations in which they were produced. Within the NHL, I found mentions of Denmark, India, Pakistan, Kashmir, Dubai, Poland and Turkey. Additionally repair notations are written in range of different languages such as Malay and Arabic that hint at other locations. This circulating cultural object is an artefact of a larger community of repair, and taken together with GSM Forum gives a sense of how widely trans-local repair knowledge extends. Is it possible to push beyond the term trans-local and assert that these are 'global' knowledge of repair? Suchman reminds us that 'locality and globalness are effects achieved in and through the discourses and practices of ICT' (2002: 140), and that what counts as 'global' should itself be an empirical question. It is clear that in the making of the NHL and the interactions on GSM Forum many local contributions are networked and interconnected across wide geographical spaces. What emerges here is something like a global knowledge of repair, where the artefacts and practices described in this chapter invite a re-specification of the global, which emerges as an incredibly seamful and non-coherent set of connections.

Conclusion

This chapter has explored the connections made by repair technicians in downtown Kampala to trans-local repair knowledge, primarily accessed via the Internet. I began by tracing distinctions between 'authorised' and independent workshops. In order to be able to restore phones to functionality, technicians draw on design and engineering knowledge about how devices are designed to operate, and how they normatively behave when working. These are foreclosed within regimes of 'authorisation,' where manufacturers restrict information about devices, making them available only to select, sanctioned workshops. Repair unfolds within these radically asymmetrical ecologies of knowledge, where independent technicians operate in the absence of information about devices, and in competition with well-resourced 'authorised' counterparts. These relationships demonstrate how the practice and possibility of repair is bound up with wider dynamics of power through the uneven distributions of technical knowledge. Active work on the part of manufacturers in maintaining and upholding these asymmetries ties repair to manufacturing centres, and undermines fixes that happen outside of corporate control. Repair is still possible for independent technicians, but the work of searching already present in repair is made more extensive, as technicians work to locate the firmware files or schematics needed to solve an already identified problem. This is an important repair story, and one that is applicable far beyond downtown Kampala, as relationships of 'authorisation' are common features of the distribution of commodity technologies in all parts of the world. Unsettled questions are raised around agency over devices, and what rights owners should have to repair their devices, set against manufacturers' intellectual property protections, within a commercial context that is temporally ordered towards the production of the new. At stake is the possibility for independent repair, which is imperilled by proprietary closures both in Southern and Northern contexts. What responsibilities should manufacturers have to contribute to ecologies of knowledge?

Within this chapter, I have used empirical cases to show how trans-local sites of repair, accessed via the Internet provide alternative venues for locating repair knowledge. 'Solutions' are produced collaboratively across geographically dispersed technician communities, and are open to all. Sites like Shrakfilestore, GSM Forum and artefacts such as the NHL are evocative of incredibly rich trans-local ecologies of repair knowledge that inform and extend technicians' practices. Using empirical fieldwork on GSM Forum, I have demonstrated how repair knowledge is generated within, and circulates through textual interaction, and the posting of augmented pictures of motherboards and schematics. Technicians collaborate by creating a thread asking for support, which is open for others to reply. The mediation of this communication on GSM Forum allows these particular instances of communication and collaboration across distance to be preserved. The individual interactions between technicians aggregate into a searchable archive of shared knowledge about repair. These are not 'stories' in the rich communicative sense that Orr intends (1996). Community norms ensure that problems are clearly framed, and replies are succinct, which helps to shape their future usefulness as 'solutions'. Nevertheless, these threads are accounts of breakdown, things that have been tried, and possible 'solutions', and they remain directed towards repair practice.

As forms of repair knowledge aggregate they begin to cohere, and responses to particular breakdowns become stabilised (to varying degrees) and known as 'solutions'. The challenge for technicians in downtown Kampala is to locate circulating solutions that address the breakdown at hand. The idea of 'online searching' glosses the iterative activity of generating solutions through queries, scanning them, understanding their relevance, and taking forward particular actions at the workbench. Searching is perhaps more accurately framed as construction or integration work that blends together trans-local suggestions for action and local knowledge, to produce a working phone. There are of course limits to what aspects of repair knowledge can travel: the richly embodied, tacit knowledge generated in interactions with devices cannot be conveyed in textual exchanges or visual notation. Working out 'solutions' at the workbench requires a high degree of bodily competence. These forms of knowledge circulate across the rich local collaborative networks within the downtown area. As technicians themselves emphasise in the interview excerpts dotted throughout the text, the importance of trans-local repair knowledge cannot be overstated. These resources extend the range and reach of repair practices in downtown Kampala. Searching is a vital strategy technicians use to manage the contingency of repair work, within sites that may be marginalised from 'authorised' knowledge and located in a geographical place that is peripheral to global circulations of technologies, tools and knowledge more widely.

Through the restoration of artefacts, repair makes a profound contribution to the stability and endurance of practices, and by extension the durability of infrastructures. Repair businesses in downtown Kampala provide appropriate and affordable services to their customers, many of whom have invested heavily in a mobile device and are keen to make the most of this expenditure. Technicians keep mobiles enrolled on telecommunications networks, so that Ugandans enjoy the benefits of telephone communication and (increasingly) mobile Internet access. Open and collaborative trans-local ecologies of knowledge contribute to the sustainability of repair as a job. Technicians rely on repair work for their livelihoods and in the case of independent workshops, getting enough money is always precarious and challenging. Technicians use their earnings to embrace the richness of life, by socialising with friends, buying a car, getting married, taking care of children, and in some cases buying land, and building or adding rooms to a house.

In aggregate, repair knowledge artefacts start to open up a particularly interesting and unique view on the wider patternings of breakdown. Repair constitutes a different way of knowing devices, that relies on, but diverges from engineering and design knowledge. The politics of who can access engineering and design knowledge remains a central issue that determines the distribution and sustainability of repair practices themselves. As we have seen, independent technicians in Kampala mitigate their peripheral position by drawing on trans-local bodies of repair knowledge made for technicians, by technicians.

Notes

1. Figures from the International Telecommunications union, generated using their Data Explorer Tool http://www.itu.int/ITU-D/ict/statistics/explorer/index.html.
2. GSM Forum is available at http://forum.gsmhosting.com/vbb/index.php.
3. Time series by country produced by the International Telecoms Union in 2013 http://www.itu.int/en/ITU-D/Statistics/Pages/stat/default.aspx.
4. The thread is available at http://forum.gsmhosting.com/vbb/f62/gsmhosting-rules-1591101/.

References

Ahmed, Syed Ishtiaque, Stephen J. Jackson, and Md. Rashidujjaman Rifat. 2015. Learning to fix: Knowledge, collaboration and mobile phone repair in Dhaka, Bangladesh. *Proceedings of the Seventh International Conference on Information and Communication Technologies and Development* 4 (1–4): 10.

Dant, Tim. 2010. The work of repair: Gesture, emotion, and sensual knowledge. *Sociological Research Online* 15 (3): 1–7.

Erickson, Thomas. 1997. Social interaction on the net: Virtual community as participatory genre. *Proceedings of the Thirtieth Hawaii International Conference on System Sciences*, 13–21.

Gartner. 2015. Gartner says smartphone sales surpassed one billion units in 2014. http://www.gartner.com/newsroom/id/2996817. Accessed 8 October 2015.

Graham, Stephen, and Nigel Thrift. 2007. Out of order: Understanding repair and maintenance. *Theory, Culture and Society* 24 (3): 1–25.

GSM Association. 2015. The mobile economy report 2015. https://www.gsma.com/mobileeconomy/global/2015/GSMA_Global_Mobile_Economy_Report_2015.pdf. Accessed 8 October 2015.

Haraway, Donna. 1988. Situated knowledges: The science question in feminism and the privilege of partial perspective. *Feminist Studies* 14 (3): 575–599.

Henke, Christopher R. 1999. The mechanics of workplace order: Toward a sociology of repair. *Berkeley Journal of Sociology* 44: 55–81.

Jackson, Steven J. 2014. Rethinking repair. In *Media Technologies: Essays on Communication, Materiality, and Society*, ed. Tarleton Gillespie, Pablo J. Boczkowski, and Kirsten A. Foot, 221–239. Cambridge: MIT Press.

Jackson, Stephen. J., Alex Pompe, and Gabriel Krieshok. 2012. Repair worlds: Maintenance, repair, and ICT for development in rural Namibia. In *Proceedings of the ACM 2012 Conference on Computer Supported Cooperative Work*, ed. ACM, 107–116.

Orr, Julian E. 1996. *Talking About Machines: An Ethnography of a Modern Job*. Ithaca: Cornell University Press.

Pollock, Neil, Robin Williams, Luciana D'Adderio, and Christine Grimm. 2009. Post local forms of repair: The (extended) situation of virtualised technical support. *Information and Organization* 19 (4): 253–276.

Sanne, Johan M. 2010. Making matters speak in railway maintenance. In *Ethnographies of Diagnostic Work: Dimensions of Transformative Practice*, ed. Monica Büscher, Dawn Goodwin, and Jessica Mesman, 54–72. Houndmills: Palgrave Macmillian.

Sheehan, Molly O. 2003. Communication networks expand. In *Vital Signs 2003*, ed. WorldWatch Institute, 60–62. http://www.worldwatch.org/brain/media/pdf/pubs/vs/2003_networks.pdf. Accessed 8 October 2015.

Star, Susan Leigh. 1995. Introduction. In *Ecologies of Knowledge: Work and Politics in Science and Technology*, ed. Susan Leigh Star, 1–38. Albany, NY: SUNY Press.

Suchman, Lucy. 2002. Practice-based design of information systems: Notes from the hyperdeveloped world. *The Information Society* 18 (2): 139–144.

———. 2007. *Human-machine Reconfigurations: Plans and Situated Actions*. Cambridge: Cambridge University Press.

Welte, Harald. 2010. Anatomy of contemporary GSM cellphone hardware. http://laforge.gnumonks.org/papers/gsm_phone-anatomy-latest.pdf. Accessed 8 October 2015.

6

The Dance of Maintenance and the Dynamics of Urban Assemblages: The Daily (Re)Assemblage of Paris Subway Signs

Jérôme Denis and David Pontille

September 7th, 2007, half past six in the morning. While the guys are finishing their cup of coffee, the sheets of paper pile up in the fax machine of the signage maintenance department. The manager takes a look at the pile, quickly sorts the sheets out, and distributes them to the two pairs of maintenance workers. Each document is a request for repairing a subway sign among the numerous ones that compose the wayfinding system. For maintenance workers, these requests are work orders that indicate the type of sign, a code that identifies the sign's emplacement within the station, and of course the station name itself.

Once the pairs of workers have received the work orders of the day, they take the signs they have to install, which are stored in a specific room of the workshop, and load them into the van. They are ready to go to the first station. Brian has just sorted out the work orders to organize

J. Denis (✉) · D. Pontille
Centre de Sociologie de l'Innovation, Centre of Sociology of Innovation, i3, CNRS, Mines ParisTech, Paris, France
e-mail: jerome.denis@mines-paristech.fr

D. Pontille
e-mail: david.pontille@mines-paristech.fr

I. Strebel et al. (eds.), *Repair Work Ethnographies*,
https://doi.org/10.1007/978-981-13-2110-8_6

interventions, and he suggests starting with the replacement of a damaged signboard in Gare de l'Est. When in the station, Brian and Jonathan go over the halls, the corridors and the platforms in search for the defective sign. They carefully look at the signs displayed in the station, especially drawing on the arrows to locate the one they are searching for. They eventually find the flawed signboard at the crossing of two corridors.

But what exactly is a flawed sign? As any wayfinding system is an arrangement of graphic components, failures mainly consist in a diverse range of visible problems concerning the material composition of signs: a battered enameled plaque, a smashed or broken PVC sheet, a ripped layer or a failing frame inside a lightbox. If some of these failures are obvious, others are subtler such as the presence of minute traces of rust or mold, the display of irrelevant or obsolete information, and even the very absence of a sign. Although different, all of these cases introduce a greater or lesser extent of disruption in the visual environment of the transportation system. In this case, the flawed sign is easily recognizable for Brian and Jonathan: its colors obviously faded out (Fig. 6.1). Jonathan easily puts it down and replaces it with the brand new one in a few minutes. From now on, the network of signs in the station is repaired. The sign does not contrast among the other ones anymore. It has taken back its part in the seamless deictic chain of references that is meant to ease riders' mobility throughout the subway spaces, and the whole city.

This short scene describes a mundane intervention that concerns an object we usually take for granted. Yet, this operation bears witness of a maintenance work that is distributed in sociomaterial practices through which some people in the transportation carrier take care of subway signs (Denis and Pontille 2014, 2015). Though sometimes prosaic, each repair operation that punctuates this maintenance work enacts the daily presence of an apparatus dedicated to the graphic ordering of an urban setting, an apparatus that plays an essential role in the heterogeneous assemblage of public transportations.

For several years now, at the crossroads of human geography and actor–network theory, ample research has contributed to redefining some of urban studies' traditional themes and to "decenter" some of its main objects and issues (Farias and Bender 2010). A large part of these analyses have been focused on the notion of assemblage, which has recently come to the fore (Brenner et al. 2011; McFarlane 2011).

Fig. 6.1 The replacement of a flawed signboard (© Photo: J. Denis and D. Pontille)

Borrowed from Deleuze and Guattari (1980), the notion allows for a "non-reductionist" approach to the city (DeLanda 2006) and is useful in describing forms of agency, which "cross the human-nonhuman divide" (Bennett 2005). The notion also highlights the heterogeneity of urban fabric as well as the circumstantial character of its transformations throughout time.

> (…) the concept of assemblage is particularly useful for grasping the spatially processual, relational and generative nature of the city, where 'generative' refers both to the momentum of historical processes and political economies and to the eventful, disruptive, atmospheric, and random juxtapositions that characterise urban space. (McFarlane 2011: 650–651)

This stream of research has called for a true study of the sociotechnical complexity of cities (Amin and Thrift 2002; Brenner 2004; Sonda et al. 2010), from the largest infrastructures (Graham and Marvin 2001; Varnelis 2009) to the most everyday objects (Molotch 2011; Molotch and Norén 2010; Watson 2014). In this broadening movement, the very place of space in the description of urban realities has been largely rethought. Where for most research in geography and sociology, physical territory represented the unquestioned starting point for any analysis (Cresswell 2004), numerous authors today insist that spatial properties be studied in their diversity as always-temporary, partial results of sociomaterial practices they cannot be separated from (Kärrholm 2007; Latham and McCormack 2004; Latour and Hermant 1998).

Wayfinding systems are paradigmatic of urban assemblages. Intimately linked to urban fabric (architecture, streets, highways, and practices themselves), they play a crucial role in the production of cities as both material and informational environments (Latour and Hermant 1998). Signs, though mundane objects, contribute to "modes of ordering" that perform and maintain "spaces of flows" (Knox et al. 2008) and are the essential components of the "machinery of placement" that equip mobility practices (Amin and Thrift 2002). They are part of the devices that "are overwhelmingly important in articulating the corporeal movements of people and their bodies (workers, migrants, refugees, tourists) via complex and multiple systems of physical transportation" (Graham and Marvin 2001: 8).

Yet, to understand such assemblages, a description of signs and their emplacement, even though obsessively detailed, does not suffice. Indeed, simply for remaining in place and thus contribute to the graphic ordering of urban settings, signs take work, and especially maintenance work (Denis and Pontille 2014). In this chapter, we propose to explore such work, which remains largely overlooked in the studies of urban assemblages. First, we will foreground the importance of maintenance in the daily existence of wayfinding systems. Second, we will investigate the day-to-day repair activities that the maintenance workers accomplish, and show that they consist mainly in reassembling operations. These operations are largely based on improvisation on the part of maintenance workers, and generally involve new material added to an initial variety of elements. The ethnographical analysis of this aspect of maintenance work makes it possible to understand that the sociomaterial heterogeneity of urban assemblages is a central issue of repair. Furthermore, taking a close-up view of reassembling operations reveals the importance of the inextricably connected operations of disassembling (see Dant in this volume). To be repaired and then reassembled, signs must sometimes be, to a degree, disordered. Repair thus consists not only in (re)producing solid assemblages which appear homogenous, but also implies that objects themselves can deal with cycles of assemblage and disassemblage. Finally, we will show that, in the case of wayfinding systems, repair interventions engage a very specific ecology of visible and invisible work (Star and Strauss 1999). As subway signs' visual qualities are heavily standardized, the erasure of all traces of repair is crucial for properly reassembling the wayfinding system. Invisibility of repair is thus not a "natural" consequence of the taken-for-grantedness of the wayfinding system as an infrastructure (Star and Ruhleder 1996), but the result of repair itself and the conditions of its success.

We will draw on an ethnography of the maintenance of the Paris subway wayfinding system that we conducted from March 2006 to March 2007. We gathered internal and external documents and we conducted in-depth interviews with designers, employees in charge of the graphic standards manuals, and employees from the signage maintenance department. We also shadowed maintenance workers during their daily rounds in stations and at their workshop, taking photographs in order to document the course of their action (Suchar 1997). These photos were not meant to provide primary materials that would be analyzed after their gathering. Rather, they were conceived of as initial analytical gestures; means to produce a sequential visual account of maintenance work (Wagner 2006), including the main operations, the gestures, and the tools used during repair interventions.

Signs and Their Maintenance: The Wayfinding Assemblage of Paris Subway

In Paris, the wayfinding system of the Régie Autonome des Transports Parisiens (RATP) has been renewed during the 1990s, and has been the object of an ambitious standardization policy that defines a set of signs, and fixes the shape and the materials for each of them (enameled sheet metal, stickers, PVC sheets, lightboxes and, most recently, screens). On the linguistic side, this normalized system of signs was conceived as a modular, hierarchized language, with numerous components that make possible a wide variety of combinations among arrows, icons, words, abbreviations, and colors. These elements were themselves standardized in an attempt to create graphic continuity from one place to another (Denis and Pontille 2014). For example, the color assigned to each subway line remains identical from one sign to another. Similarly, dark blue writing on a white background indicates the directions of the different lines, whereas white lettering on a blue background is used for the names of stations and exits. Thus, adding to its normalized material features, the wayfinding system introduces maximum graphic consistency. The signs have been imagined in terms of their reciprocal relationships, as elements forming an uninterrupted chain of references that provide riders with what its designers call an "Ariadne's thread" (Wiart et al. 1998).

Beyond their material and linguistic properties, the objects that make up the Paris wayfinding system were conceived in close relation to the subway spaces. Signs were designed to fit seamlessly into stations' architecture. For instance, their sizes are adjusted to the RATP's tiling, which is used as a grid for signs' emplacement. Moreover, the presence of signs is meant to divide each station into zones: entry halls with neighborhood and network maps; corridors with directional arrows, subway line numbers and the names of the travel directions; platforms with network maps, connections, station names, and the names of each station exit. This differentiation within sites is applied to the entire network: the same distinct zones are repeated from station to station, thus creating standard spaces despite their architectural disparities. In such a material semiotic assemblage between heterogeneous architectural features and normalized inscriptions, "the distinction between the building and its signs, between the text and the territory, becomes indistinct" (Fuller 2002: 236). Like signs and their graphic components, walls, hallways, stairwells, and platforms are essential parts of the Paris wayfinding system's assemblage.

Through such an ambitious policy, the wayfinding system has been invested as a central component of the service that the Parisian carrier provides to its customers. Beyond the genuine transportation of people from one point to another of its network, the RATP now offers to riders a full set of resources dedicated to their fluid displacements in the city. The manifold standardized subway signs are thought as a mobility utility in itself, transforming the transportation spaces into an always-available device that bears a situated intelligibility of the whole network.

Obviously, such a service cannot stand in signs "only", even though highly standardized. To be available all along the transportation networks, signs have to be watched for and taken care of. In 2000, the RATP created a maintenance department fully dedicated to the wayfinding system. At the time of our study, twenty employees were working there, taking care of all the signboards for all the transportation modes, and ensuring the wayfinding system's continuity of service. It was specified in the department's mission that they had to carry out the repairs requested by the station superintendents within 48 hours only. With a crew of six men and a woman exclusively dedicated to subway signs, maintenance work is action-packed. Four of them were intervening in the stations, repairing and replacing the boards, whereas the three others were staying at the shop, repairing old boards or manufacturing new ones. This crew was dealing with an average of 150 interventions per month, in the network of three hundred stations (and supposedly fifty thousand signs, but no one ever managed to count them).

The Dance of Maintenance

The repair interventions generally work in four steps. First of all, obviously, the flaws in the wayfinding system have to be noticed and reported. This is the responsibility of each station superintendent. Every morning, before opening the station they are in charge of, a form in the hand, they walk all around the corridors, the staircases, the halls, and the platforms, looking for any problems: graffiti on the walls, unpleasant odors, rats, furniture damage, homeless people… Signboards figure among the numerous things they have to check. Once the round is finished, they use the form they filled in to create digital requests that are automatically distributed in each specialized maintenance department.

Even if it is part of their official job assignation, station superintendents are not the only ones who can notice failures in the wayfinding system. During their interventions, maintenance workers detect problems as well, performing a supervision "on the spot", beyond the only signboard they are here to repair or replace. If they notice a new problem and have the right equipment, they generally operate right away. If they cannot, they create a request themselves, once back at the maintenance shop. Such on-the-spot notifications go beyond the formal organization of work and, when dealing with it, the maintenance workers generally consider they are doing the superintendents' job. Sometimes, after noticing a problem, they go directly to the station superintendent for a quick reminder, explaining both that they discovered a missing or damaged sign and that it should have been reported already.

Such episodes show that subway signs are not the most important matter in the superintendents' eyes. When it comes to supervision, maintenance work dedicated to the wayfinding system is framed in the maintenance of the subway station as a whole. Furthermore, the fact that the superintendents sometimes "miss" what the maintenance workers consider as noticeable failures foregrounds the difficulties such notifications represent. Flaws in the wayfinding system are by no means self-evident and their mere identification requires specific competencies, a "professional vision" (Goodwin 1994) through which maintenance workers articulate their ability to observe tiny transformations in their environment and their sound knowledge of the standards (that is the way signs should be in a "normal" situation).

When they proceed from the station superintendent, the requests start another sequence. Distributed to the maintenance workers, these requests describe some problems in the wayfinding system that need to be confirmed on the spot so as to be characterized in more details. Although diagnostic is a crucial operation in many repair activities (Bovet and Strebel in this volume; Dant 2008; Orr 1996; Sanne 2010), it is not at stake in the maintenance work of subway signs. The workers simply confirm or invalidate the relevance of the problem described in the work orders. To do so, they go into the station a first time and take a decision in front of the sign. In some cases, the request is clearly not addressed to the right department. For instance, whereas a "scratchiti" so deeply damages an enameled plaque that it has to be replaced by the maintenance workers, a graffiti merely painted on a sign should be removed by the cleaning department. If the failure is confirmed, the repair may be done on the spot, though it rarely happens. In most cases, the repair of the wayfinding system goes through the replacement of signs. Therefore, the first run into the station generally implies taking measurements of the damaged board. To do so, the maintenance workers complete their initial work order, confirming the type of sign (PVC or enameled), indicating its precise dimensions, and sometimes drawing its graphic composition and content.

As the form is then passed on to the maintenance department, it is the starting point for a two-step production process. Either the signboard can be directly repaired at the workshop, or a new enameled one has to be ordered from the manufacturers, based on the measurements reported in the form. But nine weeks are needed for the production and the delivery of such a sign. In order to respect the mandatory 48-hour delay, a local team makes a temporary PVC signboard at the shop. Except in particular cases, the next day this temporary sign is put up by the maintenance workers, which closes the case in the information system at the end of their round. At this time, from the point of view of the standardized policy we described above, the wayfinding system is thus considered temporarily repaired. When the final signboard is delivered to the maintenance shop, a new work order is edited and another pair of maintenance workers (sometimes the same ones) returns to the station and proceed in the replacement of the PVC one.[1]

Thus repairing the wayfinding system of Paris subway cannot be summarized as the automatic replacement of flawed signs with new ones found in a warehouse stock. Rather, it is a process punctuated by daily operations that put things back to order into a network of signs never repaired once and for all. Regular interventions set the rhythm of what we propose to call an endless "dance of maintenance", to echo Pickering (1995), who spoke of "the dance of agency." Such a dance is made up of a permanent attention to subway signs, small mundane gestures of repair, and replacement operations that oscillate between provisory and definitive states.

But what do these operations consist in precisely? What does repairing the assemblage of the wayfinding system actually mean?

Reassembling

Let us first take the example of an intervention carried out by a pair of maintenance workers (Michael and Steven) who were asked to replace a damaged PVC sheet inside a lightbox.

> On the platform, Michael sets the signboard down. Balanced on the stepladder, he loosens the screws, opens the lightbox and slides out the PVC sheet (see Fig. 6.2). The plastic layer covering the words is partly ripped. The extent of the damage did not stand out when it was in the lightbox: the fact that this damage was signaled shows the degree of rigor in keeping the wayfinding system in top-notch shape. During this time, Steven has remained at the bottom of the ladder in order to take the broken sign from Michael. Before continuing the intervention, Steven holds the old and new sheets up to one another, making sure that they are the same size. Without needing a ruler, he confirms that they are indeed identical in dimension. Michael then places the new sheet in the lightbox.

Fig. 6.2 Slides out the PVC sheet (© Photo: J. Denis and D. Pontille)

However, he is not completely satisfied: despite the metal grips that he uses to put it in place, the sign is not completely sturdy. Michael could stop there and close the lightbox; after all, both signs are identical and the

> previous one was hardly sturdier. But there is a risk that it will fall out of the box, and Michael prefers to avoid this possibility. As he explained to us, if the sign were to fall on a rider's head, he and Steven would be held responsible. He wants to find a solution, even if it takes him a few minutes longer.
>
> Without discussion, Steven goes back up to the station entrance hall. When he returns, he has newspaper in his hands, which he folds into small, thick squares. Each square is modeled after several folds. Michael positions the squares one by one between each grip and the PVC sheet in the lightbox (see Fig. 6.3). The sign is now sturdy and adjusted. The box can be closed again. Once this step is complete, Michael concludes: "Good as new! Now it looks good and it's better for the riders". (July 9th, 2007, Field notes)

This sequence reveals certain particularly interesting aspects of repair. First, it underlines a well-known dimension: the importance of improvisation and "bricolage" in the completion of repair and maintenance operations (Dant 2010; Henke 2000; Schubert in this volume). By definition, repair is made of constant surprises and adjustments that go beyond all attempts to rigorously plan things (Orr 1996). This sequence also shows, though, that improvisation is not exactly what is at stake here. Maintenance workers seem to know what to do. Not only do they not discuss with one another in order to find a solution

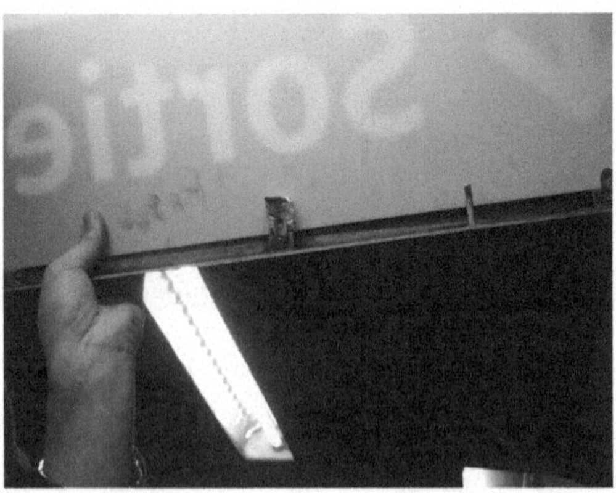

Fig. 6.3 Fitting the sign into the lightbox (© Photo: J. Denis and D. Pontille)

to a problem indirectly affecting rider safety, but they also coordinate their respective activities fluidly. Michael had barely noticed that the sign was unstable when Steven left to find the newspaper. Workers deal with routine problems from which they easily extricate themselves, no equivocation needed. And yet, putting newspaper in the box to fit the PVC sheet is a kind of bricolage. Newspaper is a material that is not part of the standard criteria for lightbox assemblage. Moreover, though routine, its use remains invisible once the lightbox has been closed.

Throughout other interventions, we observed the introduction of other, much more unexpected elements whose invisibility was not evident. Improvisation also plays an important role in these examples.

This is what happened when Brian and Jonathan attempted to replace a sign that was on the verge of falling down. During an initial station intervention we were not able to observe, two repairmen found themselves in a surprising situation. With abundant detail and no shortage of humorous remarks, Brian and Jonathan explained to us that the sign identified as flawed in the work order had come halfway off the wall. In trying to remove it entirely, they discovered that it had not been placed in its usual frame, but was resting on a frame made of rotting wood. They ended up removing the whole thing, and decided to come back to the station in order to attach it correctly.

Two days later, we shadowed them as they prepared to put the sign back in place. That day, before leaving the maintenance department, Brian asked his colleague for two metal brackets. Once in the station, the team prepared the wall, removing the old screws, drilling holes, and filling them with glue cement. Despite a few problems with their electrical tools, they finally fixed the brackets and put the signboard up. At the end of this series of operations, while we were expecting the maintenance workers to leave the station in their van, an interesting thing occurred.

Brian stops in front of the sign. He seems unsatisfied. Since he riveted the signboard to the two horizontal brackets, two empty holes, without rivets, remain visible on both sides of the board. He points to them, upset. In the terms of standards, the board is not properly put up, and he's afraid that the superintendent would not notice it has been repaired. There's a chance he would not acknowledge that the job has been done. Brian decides to put a screw in each of these empty holes, but it's even worse: the screws do not hold in place. Finally, he begins to clean the sign, which had gradually gotten covered with dust, using a cloth and the sleeve of his sweater (see Fig. 6.4). The sign sparkles, as if it had just come

Fig. 6.4 Shining the sign (© Photo: J. Denis and D. Pontille)

from the workshop. Brian smiles: "Here it is, a brand new sign!", adding "The superintendent should not be tempted to look at it too closely". (September 7th, 2007, Field notes)

These two sequences show that reassemblage is more than a simple re-establishment of already-present elements. Much on the contrary, what our observation sheds light on is that, during repair, the boundaries between the signs and their environment are neither frozen in place nor sealed or closed. While the squares of newspaper were added to hold the PVC sheet in place in the first sequence, in the second, glue, extra screws, and, above all, metal brackets specially created for the task made it possible to reinforce the sign to the wall. During their interventions, the maintenance workers constantly consider different material sites, strengthen their composition according to their own criteria, and try to make them hold as a coherent assemblage as best they can. Instead of encountering objects with stabilized boundaries, such as a wall and a sign, they are immersed in a material flux made of multiple layers. In others words, they deal with a dense ecology of materials (Bennett 2004; Denis and Pontille 2015; Ingold 2007).

The wayfinding system repair involves thus reassembling operations that not only move the borders between the assembled elements, but also sometimes transform walls and ceilings just as much as the signs themselves. The second sequence is a particularly good illustration of this. Glue, screws, and metal brackets are not just items added between a solid wall and a sign in good condition, though detached from the wall. The discovery of a wooden frame damaged during a previous intervention, then the creation of specially adjusted metal brackets show above all that the previous sign's assemblage was incomplete. Its visible side, displayed for the riders to see, was in fine condition, but its reverse side, initially made of metal so that the sign could be attached and adjusted to the wall with rivets, had disappeared. This missing half of the sign is compensated for by the workers; during the repair operations, they create their own version of this other half. By the end of the intervention, the back half of the sign is supposed to resemble the other "whole" (or supposedly whole) signboards in the wayfinding system.

This new back half of the sign is the result of an assemblage, some materials of which are different from those of the "whole" signs.

The two sequences make it possible to understand the relationships between order and disorder in repair operations, and also shed light on the ecology of the visible and the invisible on which these relationships are based. The repair practices we observed are oriented toward the restoration of a normalized situation, that is, a situation that meets standardized criteria. The maintenance workers strive to get things back to order. In the second sequence, Brian's last gesture and remark show that such "normalization" is clearly a matter of visibility and invisibility. To be considered as repaired, the signboard has to seem brand new. Such visible "newness" draws on two kinds of erasure: the erasure of the "out of order" situation and the erasure of the maintenance intervention itself and all the traces it may leave on the board.

The question of visibility and invisibility is also present in the first sequence, but it is organized differently. While the workers in the second sequence shine the repaired signboard in order to mask the aesthetic imperfections (the holes) resulting from its being incomplete, in the first sequence the workers make sure the PVC sheet is firmly attached in the lightbox, adding elements that will remain invisible. Thus, here, far from being limited to issues of recognition of workers within the formal representations of an activity (Bowker et al. 1995; Star and Strauss 1999; Suchman 1995), the ecology of the visible and the invisible is very directly concerned with objects and the unique assemblage they are part of. In one case, the pieced-together assemblage is hidden inside the lightbox itself, and, in the other, it is masked by the excessive shine of the enameled signboard.

The relationships between visible order and disorder made invisible are significantly different depending on who the intervention is aimed at. What these two sequences show is that repair engages not only the riders toward whom the brand new signboard is displayed without a trace of additional pieces (first sequence), but also the station superintendent who asked for the signboards to be repaired (second sequence). As we have seen, cleaning the sign until it shines is a means to avoid the superintendent's close inspection, which would mean he might see the unorthodox solution the workers came up with to fix it. From one situation to the next, that which is masked and that which is made visible is neither aimed at the same people, nor for the same reasons. Worker responsibility toward rider safety is at stake in the first sequence; the signboards' appearance as objects in keeping with the aesthetic standards of the wayfinding system is that which orients the workers' actions in the second.[2]

Finally, these two sequences remind us that repair is not merely re-assemblage. Repairing a signboard, whatever its materials, requires dynamic sequences which invariably include disassemblage. Disassembling and reassembling are the two essential steps in maintenance work. This is very clear at the beginning of the first sequence, when the damaged PVC sheet was removed, but it is also true for the sign in the second sequence, which was on the verge of falling, and thus literally about to "detach" itself from the wayfinding system. In neither of the two cases did the disassemblage process require complex work: in the first sequence, Michael removed the PVC sheet from its frame without much effort, and, in the second sequence, the signboard seemed to break away from the wall "all by itself." During other interventions, though, things did appear more difficult. The disassemblage operation can be, at other moments, a more significant part of repair.

This was the case when we accompanied Brian and Jonathan on an intervention to replace a PVC sheet inside a lightbox because the information displayed on the sign was obsolete.

> The first step of the intervention consists in opening the lightbox in order to take its front side down. This is a delicate operation: perched on the top of a stepladder, Jonathan opens the box and puts his hands behind the frame, trying not to get burned by the fluorescent lamp and trying not to break the whole thing. Once the piece is detached and put on the floor, the next step is quite long. In order to take out the sheet itself, Brian and Jonathan have to remove manually all the sixteen small screws that hold the plaque to the metal frame (see Fig. 6.5). Once it's done, they put up the new plaque very carefully and screw back the frame. Finally, they put back the piece in front of the lightbox. After that, Jonathan tells us: "This makes no sense, 16 screws just to get at this sheet. All we'd need is a little trap along here and we could easily get the sign out without even having to open the box up, without needing to take anything out… But they don't think about that. They don't think about us." (July 4th, 2007, Field notes)

Fig. 6.5 Removing screws (© Photo: J. Denis and D. Pontille)

In this sequence, screws are dealt with once again. But when in the earlier sequence, they were briefly considered as a way of masking holes in the enamel sheet metal and making the signboard half appear complete and "closed", here, screws are firmly attached to the metal frame. They are a constraint in the disassembling process. There are many of them, and they guarantee both the solidity of the signboard and the safety of the riders. When the PVC sheet was unsturdy in its lightbox in the first sequence, screws were sorely lacking. But in this last sequence, they represent a true obstacle to quick signboard repair.

This sequence does not only shed light on the importance of disassemblage operations at the heart of repair, though. It also underlines the tension that may exist between the need to produce solid, lasting assemblages, either in the name of a placing signboards everywhere in order to serve rider mobility, or in the name of rider safety, and the importance of designing objects that can still be easily disassembled, which is an essential quality when it comes to repair. The attention paid to disassemblage operations is a way of not remaining limited to the heterogeneity of the elements that make up the wayfinding system in particular, and sociotechnical devices in general. It involves taking into consideration object agency in maintenance work, paying special attention to the material conditions of their repairability (de Laet and Mol 2000; Denis and Pontille 2015).

Conclusion

In studying urban assemblages through the lens of repair ethnography, we have attempted to pursue recent reflections which have demonstrated the interest of a decentered analysis of the urban realities linked to the traditional investigations of urban studies. We have shown that studying the organization of maintenance work, and observing the conditions under which repair is accomplished as closely as possible allows to go beyond the mere observation that cities are sociomaterially heterogeneous. It notably makes possible to apprehend urban assemblages in their daily dynamics and to investigate the ecology of visible and invisible in which their repair takes place.

The Dynamics of Assemblages

To paraphrase Haraway (2003), we can therefore say that repair operations show that urban assemblages can be understood as "active verbs". There are two ways of conceiving of this. First, we can insist on the fact that urban assemblages are moving wholes, with always changing borders and components (Brand 1994; Edensor 2012; Jones and Yarrow 2013). We observed notably that the assemblage of the Paris subway wayfinding system is not only made up of permanent signboards with varying shapes and functions, articulated in a complex modular system, but it is also composed of temporary signboards, hung daily to make up for the lack of full repair, given the slow rhythm of production of the enamel signboards. These temporary signs make possible a kind of relative permanence for the system as a whole. Their presence though keeps the Paris wayfinding system from remaining entirely identical from one day to the next.

The second, complementary way to describe urban assemblages as "active verbs" is to highlight their precariousness and vulnerability. Heterogeneous parts do not hold together once and for all after they have been designed or installed. Rather, they are the product of constantly changing relationships (Strebel 2011; Edensor 2012). In the case of the wayfinding system, signboards, whether "definitive" or "temporary," are always fragile, subject to wear and tear, breakage, even theft (Denis and Pontille 2014). The transformations we have stressed here are symptoms of the unceasing activities which guarantee an assembled whole. This is what we have attempted to illustrate in presenting the dynamics of assemblage and re-assemblage which characterizes the repair of the Paris wayfinding system. Beyond, or rather, short of, the master narratives describing the terrible risks of disaster that each city faces, this unfailing maintenance process reminds us that cities are repaired daily, and incessantly. Rather than focusing the analysis on urban settings and, in particular, on their infrastructures, in pitting normality and crisis or functioning and breakdown against one another:

> we need to be especially mindful of the continuous, invisible work necessary to bring about infrastructural circulation even when infrastructural assemblages are working "normally." (Graham 2010: 19)

Urban assemblages are thus constantly going through assembling and reassembling operations, and these guarantee their permanence and their "normality". This is why maintenance can be understood as a dance: a series of coordinated movements that punctuate the life of the objects and infrastructures, partly through planned and repeated occurrences, partly through improvisations of all kinds.

Performing In/Visibilities

Finally, the case of the wayfinding system daily maintenance raises an important aspect of repair: its inscription in a very specific ecology of visible and invisible (Star and Strauss 1999). We have seen that part of repair activities makes visible certain characteristics of objects which, in a way, prove that they are back in working order. This visibilization relies on the erasure of the bodies of the workers, of the traces of their interventions, and the removal of the ingredients added to the assemblage for the sake of repair. The ethnography of repair therefore lets us "surface" (Star 1999) not only the invisible work of maintenance workers, but also the conditions by which it is erased, and the dynamic ecology of the visible and the invisible that fuels the daily existence of certain urban assemblages.

In this respect, the case of wayfinding systems seems very particular. Indeed, such apparatuses mainly operate on a visible basis. The intelligibility of the transportation network is performed by the very presence of the signs to the eyes of all riders. In the case of Paris transportation systems, we saw that, through the standardization of the shape, content, and emplacement of subway signs, specific visual qualities have been invested as key ordering operators. Hence, what repair operations mainly do in such cases is to restore the signboards' proper visibility, which is considered as the very condition of the service the wayfinding system is aimed at providing. To perform such a restoration requires making all things that are not initially designed as part of the system invisible, from additional materials to traces of repair.

The flaws themselves, of course, are made invisible in such a process. Yet, to be repaired, they have to be noticed, that is somehow to be visible. This is an important aspect of the ecology of visible and invisible in which the maintenance work dedicated to such highly visible a thing as wayfinding systems is engaged: the visibility of the signs' standardized and harmonized features is relational. It is performed from one repair operation to another, during which failures are visible for maintenance workers and remain, at least they hope, invisible from the riders' point of view. More generally, such an ecology seems to be specific of most of maintenance work, which mainly consists in accomplishing a flow of small interventions, in contrast with major breakdowns that require more substantial repair. In science and technology studies, as in phenomenology, breakdowns and repair situations are usually described as occasions for bringing hitherto unnoticed aspects of the world to light. Infrastructures for instance are ordinarily taken for granted, until they collapse and we suddenly understand and experience their importance and their vulnerability (Graham 2010; Star and Ruhleder 1996;

Star 1999). But if this is true from the point of view of the everyday users, it is not from the one of maintenance workers, who deal on a daily basis with the fragility of infrastructures. Maintenance work draws a less contrasted situation where the difference between functioning infrastructures or technologies and broken ones is not binary, but relational (Denis and Pontille 2017). What counts here is not the visibility or invisibility of failures, repair traces, or transformations in general, but the distribution of people who are supposed to notice them or not.

Notes

1. In the case of signs made of a PVC sheet framed inside a lightbox, the version produced at the shop is the final one.
2. In his study of the mutable stone of St Ann's Church, Edensor showed that expectations surrounding the decisions at the heart of repair are multiple, vary throughout time, and may be hotly debated: "Decisions about repair, aesthetic appearance, historical worth, architectural and heritage value may persist as orthodoxy for a period of time or they may be hotly contested. [...] Runcorn stone becomes a widely popular building material across north-west England for a time before becoming unsuitable for heavily polluted industrial settings; sandblasting prevails as a repair technique but is discredited; biofilms are left to grow because they are not currently considered to impair aesthetic appreciation" (Edensor 2011: 249).

References

Amin, Ash, and Nigel Thrift (eds.). 2002. *Cities: Reimagining the Urban*. Cambridge: Polity Press.

Bennett, Jane. 2004. The force of things: Steps toward an ecology of matter. *Political Theory* 32 (3): 347–372.

———. 2005. The agency of assemblages and the North American blackout. *Public Culture* 17 (3): 445–465.

Bowker, Geoffrey C., Stefan Timmermans, and Susan Leigh Star. 1995. Infrastructure and organizational transformation: Classifying nurses' work.

In *Information Technology and Changes in Organizational Work*, ed. Wanda J. Orlikowski, Geoff Walsham, Matthew R. Jones, and Janice I. Degross, 344–370. London: Chapman and Hall.

Brand, Stewart. 1994. *How Buildings Learn: What Happens After They're Built*. New York: Viking Penguin.

Brenner, Neil. 2004. *New State Spaces: Urban Governance and the Rescaling of Statehood*. Oxford: Oxford University Press.

Brenner, Neil, David J. Madden, and David Wachsmuth. 2011. Assemblage urbanism and the challenges of critical urban theory. *City* 15 (2): 225–240.

Cresswell, Tim. 2004. *Place: A Short Introduction*. Malden, MA: Blackwell.

Dant, Tim. 2008. The 'pragmatic' of material interaction. *Journal of Consumer Culture* 8 (1): 11–33.

———. 2010. The work of repair: Gesture, emotion, and sensual knowledge. *Sociological Research Online* 15 (3): 1–7.

de Laet, Marianne, and Annemarie Mol. 2000. The Zimbabwe bush pump: Mechanics of a fluid technology. *Social Studies of Science* 30 (2): 225–263.

DeLanda, Manuel. 2006. *A New Philosophy of Society: Assemblage Theory and Social Complexity*. New York and London: Continuum.

Deleuze, Gilles, and Félix Guattari. 1980. *Milles Plateaux*. Paris: Ed. de Minuit.

Denis, Jérôme, and David Pontille. 2014. Maintenance work and the performativity of urban inscriptions: The case of Paris subway signs. *Environment and Planning D: Society and Space* 32 (3): 404–416.

———. 2015. Material ordering and the care of things. *Science, Technology, & Human Values* 40 (3): 338–367.

———. 2017. Beyond breakdown: Exploring regimes of maintenance. *Continent* 6 (1): 13–17.

Edensor, Tim. 2011. Entangled agencies, material networks and repair in a building assemblage: The mutable stone of St Ann's Church, Manchester. *Transactions of the Institute of British Geographers* 36 (2): 238–252.

———. 2012. Vital urban materiality and its multiple absences: The building stone of central Manchester. *Cultural Geographies* 20 (4): 447–465.

Farias, Ignacio, and Thomas Bender. 2010. *Urban Assemblages. How Actor-Network Changes Urban Studies*. New York: Routledge.

Fuller, Gillian. 2002. The arrow–directional semiotics: Wayfinding in transit. *Social Semiotics* 12 (3): 231–244.

Goodwin, Charles. 1994. Professional vision. *American Anthropologist* 96 (3): 606–633.

Graham, Stephen. 2010. When infrastructures fail. In *Disrupted Cities*, 1–26. New York: Routledge.

Graham, Steve, and Simon Marvin. 2001. *Splintering Urbanism: Networked Infrastructures, Technological Mobilities and the Urban Condition*. London: Routledge.

Haraway, Donna Jeanne. 2003. *The Companion Species Manifesto: Dogs, People, and Significant Otherness*. Chicago: Prickly Paradigm Press.

Henke, Christopher R. 2000. The mechanics of workplace order: Toward a sociology of repair. *Berkeley Journal of Sociology* 44: 55–81.

Ingold, Tim. 2007. Materials against materiality. *Archaeological Dialogues* 14 (1): 1–16.

Jones, Siân, and Thomas Yarrow. 2013. Crafting authenticity: An ethnography of conservation practice. *Journal of Material Culture* 18 (1): 3–26.

Kärrholm, Mattias. 2007. The materiality of territorial production: A conceptual discussion of territoriality, materiality, and the everyday life of public space. *Space and Culture* 10 (4): 437–453.

Knox, Hannah, Damian O'Doherty, Theo Vurdubakis, and Chris Westrup. 2008. Enacting airports: Space, movement and modes of ordering. *Organization* 15 (6): 869–888.

Latham, Alan, and Derek P. McCormack. 2004. Moving cities: Rethinking the materialities of urban geographies. *Progress in Human Geography* 28 (6): 701–724.

Latour, Bruno, and Émilie Hermant. 1998. *Paris Ville Invisible.* Paris: La Découverte.

McFarlane, Colin. 2011. The city as assemblage: Dwelling and urban space. *Environment and Planning D: Society and Space* 29 (4): 649–671.

Molotch, Harvey L. 2011. Objects in the city. In *The New Blackwell Companion to the City*, ed. Gary Bridge and Sophie Watson, 66–78. London: Blackwell.

Molotch, Harvey L., and Laura Norén. 2010. *Toilet: Public Restrooms and the Politics of Sharing*. New York: New York University Press.

Orr, Julian E. 1996. *Talking About Machines: An Ethnography of a Modern Job*. New York: Cornell University Press.

Pickering, Andrew. 1995. *The Mangle of Practice: Time, Agency, and Science.* Chicago: University of Chicago Press.

Sanne, Johan M. 2010. Making matters speak in railway maintenance. In *Ethnographies of Diagnostic Work: Dimensions of Transformative Practice*, ed. Monica Büscher, Dawn Goodwin, and Jessica Mesman, 54–72. Houndmills: Palgrave Macmillian.

Sonda, Giovanna, Claudio Coletta, and Francesco Gabbi (eds.). 2010. *Urban Plots, Organizing Cities*. Farnham: Ashgate.

Star, Susan Leigh. 1999. The ethnography of infrastructure. *American Behavioral Scientist* 43 (3): 377–391.

Star, Susan Leigh, and Karen Ruhleder. 1996. Steps toward an ecology of infrastructure: Design and access for large information spaces. *Information Systems Research* 7 (1): 111–134.

Star, Susan Leigh, and Anselm Strauss. 1999. Layers of silence, arenas of voice: The ecology of visible and invisible work. *Computer Supported Cooperative Work (CSCW)* 8 (1): 9–30.

Strebel, Ignaz. 2011. The living building: Towards a geography of maintenance work. *Social and Cultural Geography* 12 (3): 243–262.

Suchar, Charles S. 1997. Grounding visual sociology research in shooting scripts. *Qualitative Sociology* 20 (1): 33–55.

Suchman, Lucy. 1995. Making work visible. *Communications of the ACM* 38 (9): 56–64.

Varnelis, Kazys (ed.). 2009. *The Infrastructural City*. New York: Actar.

Wagner, Jon. 2006. Visible materials, visualised theory and images of social research. *Visual Studies* 21 (1): 55–69.

Watson, Sophie. 2014. Mundane objects in the city: Laundry practices and the making and remaking of public/private sociality and space in London and New York. *Urban Studies* 52 (5): 876–890.

Wiart, Alain, Ariane Le Roux, and Marc Lomazzi. 1998. Signalétique, le nouveau fil d'Ariane. *La Vie du rail et des transports* 57: 30–35.

7

Inquiring and Experimenting with Urban Ecology: Pragmatist Lessons from Public Bicycle Repair

Martin Tironi

Introduction

> I always say I like rummaging around in the rubbish, because when the same piece fails repeatedly, you realize something is going on [...]. That's why it's a good idea to keep an eye on the broken parts and breakdowns that occur [...]. For example, when a frame is bent, we soon realize that the bicycle was vandalized, or that someone tried to steal it. (Interview with Velib' maintenance agent)

This explanation, from one of the Vélib' maintenance agents in Paris,[1] reveals a key element in the argument developed in this paper. Its aim is to show how capacity for inquiry, innovation and experimentation emerges as maintenance and repair agents deal with broken down bicycles. This argument may seem counter-intuitive, given that acts of maintenance and repair normally have an inferior status in urban life. They are routinely considered as less noble than the practices of the so-called

M. Tironi (✉)
Pontificia Universidad Católica de Chile, School of Design, Santiago, Chile

I. Strebel et al. (eds.), *Repair Work Ethnographies*,
https://doi.org/10.1007/978-981-13-2110-8_7

'creative class' (Florida 2002), which emerge hand-in-hand with the Smart City boom (Tironi and Valderrama 2018). Given the supposedly repetitive and grubby nature of the work, generally carried out by 'labouring class', these acts are often overlooked when it comes to analysing sources of innovation in a city. From this perspective, maintenance does not manufacture anything; it does not add value to the urban experience and is instead seen as an automated task that has very little to do with innovation.

However, by examining the case of the maintenance of the Parisian public bicycle scheme (Vélib), this paper aims to problematise this vision, demonstrating how maintenance and repair can be conceptualised as forms of urban innovation, and the city as a place of inquiry and experimentation that is constantly open to frictions and inventive encounters. Although the Vélib' programme can be considered as a relatively solid *black boxing*,[2] we will see varying degrees of infrastructural flexibility in the practices of the maintenance and repair agents, and how the inquiry work of these agents comes to play a central role in the ways in which the public bicycle infrastructure is adapted.

As indicated by various authors (Star 1999; Hannam et al. 2006; Graham and Thrift 2007; Denis and Pontille 2010, as well as their chapter in this volume), carefully researching the sociotechnical organisations behind the mobility and circulation of objects and persons is an important, even political task. As indicated by Latour (2009: 9) "we can only talk about mobility because there is an immobile infrastructure. When the mobility of an element increases, the immobility of the infrastructure increases accordingly." By differentiating research approaches, which are focused on the so-called *mobility environments*,[3] we aim to show other mobility categories, which could be described as '*the immobiles of mobility*' (Hannam et al. 2006; Latour 2009). Instead of taking the movement of people and objects as an inevitable element, we will examine how this movement is enabled by those responsible for the scheme's maintenance.

We will examine the impact of breakdowns on the programme and the forms of knowledge generated by these situations (Dant in this volume; Henke 2000; Denis and Pontille 2010). We will focus on the skills developed by the maintenance agents to identify breakdowns and recognise their causes, changes and impact, in an ongoing effort to find the best way of dealing with a service problem. Breakdowns are described as elusive, subtle and sometimes invisible, objects as their effect on the bicycles is never uniform or stable; their limits are unclear and it is difficult to know when they may appear or disappear. We will show that the uncertainty produced by breakdowns requires agents to develop experimental operations in which maintenance work is not confined to restoring an object to its original state, but instead involves active operations of knowledge production. The maintenance agents are not simply managing a transport system; they are active *mediators* and versatile experimenters of an infrastructure that is continuously redefining its relationship with the urban ecology (Latour and Hermant 1998). Through their daily interventions in the city, agents actively identify and solve problems, while formulating definitions of what a Vélib' 'user' is and should be.

This paper will show the centrality of the inquiry operations used by maintenance agents to identify, clarify and resolve problems. Maintenance agents are continuously faced with indeterminate situations. According to Dewey (1938), indeterminate situations are inherent in all processes of inquiry, when people build their knowledge on the basis of situations declared problematic. Inquiry originates from a doubt associated with the indeterminate nature of the situation, in response to which one tries to determine which aspects are at stake in the situation, reconsidering issues that were previously taken for granted (Dewey 1938). This paper will show that the 'indeterminate situations' are neither a handicap nor an obstacle to investigation. On the contrary, and as Stark noted with reference to Dewey's work:

> destabilising situations are the special moments through which the inquirer discovers what is at stake, because it is in these situations that the actors themselves come to realise what they had previously considered as acquired. (Stark 2009: 32)

We will demonstrate that agents' empirical experience—faced with an urban environment that must be prepared for the scheme to work well—is littered with these indeterminate situations and that their actions are largely guided by the principle of inquiry and exploration.[4] Vélib's handling and resolution of practical problems tests a type of inquiry *ethic*, which involves situated and embedded knowledge of the situation specific to maintenance and repair. In this sense, here we can understand repair as socio-material and relational (Callon 1986; Henke in this volume; Denis and Pontille 2010) to show how inquiry carried out by agents in response to problematic situations requires not only situated and distributed knowledge of human and non-human elements (discourses, materialities, relationships, etc.), but also displays a capacity for innovation in order to redefine and renegotiate certain social and technical elements of the service. That is to say, agents are not only able to deal with problematic situations, they also become active mediators between the conceptual universe and the service user's world by engaging in inquiry.

Two ethnographic data collection techniques were used in our study carried out between February 2010 and August 2011. First, we carried out observations by accompanying actors from the company directly in charge of Vélib' maintenance, namely, heads of sector, and bicycle technicians or maintenance agents during their daily activities. Maintenance agents were accompanied systematically during full working days (from 6:30 to 14:30). Using field notes, photographs,[5] and audio recordings, we documented the various operations and problems these agents encounter every day. Using the shadowing method (Grosjean and Thibaud 2001), we reviewed these observations for what the actors describe as their own interventions in progress, their work sequences, rhythms, relationships with clients, and essential tools and techniques. Furthermore, we conducted in-depth interviews with various key figures within the Vélib' system (director, head of workshops and inventories, head of operations), who do not necessarily have direct contact with the infrastructure in the field, but who play a fundamental role in the organisation of the system.

Inquiry into 'Natural Breakdowns'

The bicycles used in public bicycle systems were designed from the outset with a clear emphasis on materials that would resist inclement weather and possible acts of vandalism. The weight of the bicycles (21.5

kilos) and the decision to conceal all of the different cables (brakes, gears and lights) within the frame were the result of the need to design sturdy bicycles that would withstand intensive use by the general public.

Despite these precautions, the Vélib' bicycles have experienced significant wear and tear since their launch, in 2007: punctures, worn down brakes, rusted frame, broken chains and flaking paintwork. Contrary to predictions, the on-street maintenance and workshop repairs (in cases where bicycles cannot be repaired in situ) have come to have a far greater importance than originally anticipated, which has prompted a restructuring of the service on an organisational level, including, for example, increasing the number of maintenance agents, developing public awareness campaigns and strengthening certain bicycle parts.

But learning to cope with the world of bicycle faults and failures takes on a radical character when observed at ground level, in the situated practices of the maintenance agents. As one of the agents explained, "we live in a city plagued by faults; our job is to chase down the faults and make sure the system works properly." The agents practically deal with the city and the transport technology from a perspective of dysfunctionality, of damaged parts and materials. For them, the faults are not anomalies (as they may be for a client), but an omnipresent and natural universe, which they must learn to understand and coexist with.

Thus, one of the strategies developed by agents to navigate through this world of anomalies, faults and dysfunctions is to devise categories and distinctions. In turn, the maintenance agents make a distinction between 'natural breakdowns'—resulting from wear and tear through use—and breakdowns caused by vandalism, which we refer to as 'social breakdowns'. This second category of faults, resulting from antisocial acts, has caused the most severe of problems for the service.

> A bicycle is used 80 times a day, causing wear and tear and natural failure of various parts of the bicycle. So numerous parts have to be replaced due to natural wear and tear. But if you add in vandalism, there comes a time when you can no longer balance the maintenance repairs, the natural product of the bicycle's wear and tear, with the repairs to breakdowns caused by vandalism. (Interview with the Director of the Vélib' programme)

We will first focus on the agents' activities in dealing with breakdowns they classify as 'natural'. According to the terms of the contract between the Council of Paris and the company in charge of maintenance (who operate the service), regardless of whether the damage or breakdown is natural or not, bicycles must be repaired within 48 hours (as in the case of the city's subway signs, see Denis and Pontille in this volume). This requirement poses a real challenge to the service's maintenance agents; they do not perceive a Vélib' bicycle as a passive and mobile system as most users do, but as a potentially dangerous object that may pose a risk to clients, incite vandalism, and very often, cause problems for agents during their interventions. Although the fluidity of bicycles between stations enables the service to function, this fluidity also creates the biggest problems for agents. Freedom of movement of these objects is a permanent problem for maintenance agent.

> On the metro or bus, I can anticipate your journey. However, with the freedom offered by the Vélib' system, I cannot tell which way you will go. We are victims of the freedom we give users, which makes maintenance all the more complicated. (Interview with the Head of Operations)

Bicycles requiring maintenance are permanently on the move, they do not 'communicate' their breakdowns or problems; neither do the users, who will very often continue riding a faulty bicycle. Two years after the Vélib' scheme was launched a practice emerged whereby users would rotate the saddle backwards to indicate that a bicycle was faulty. The 'Vélib Users Committee', set up to improve the service by the City of Paris and JCDecaux in March 2010,[6] proposed the idea of user-actors to establish the bases for client collaboration in the system's maintenance. However, for those working in field on a daily basis, users' technical solidarity was more anecdotal than real.

> I think that this practice of rotating the saddles of faulty bicycles is misleading, because often users just do it to reserve a bicycle for later. If this practice were 100% reliable, then we would just have to check the bicycles with the saddles pointing backwards but unfortunately this is not the case. (Interview with the Sector Head)

Objects in movement, bicycles and users that do not report problems—and may even feign problems to take advantage of the system—make the question of having legitimate 'messengers' of breakdowns a complex issue in terms of maintenance. Breakdowns create situations that are difficult to understand; situations with no inventory of intrinsic properties, hybrid entities that are both human and material, fixed and mobile, technical and social, visible and invisible.

Making Invisible Breakdowns 'Visible'

Vélib's senior management and maintenance agents describe the problem of bicycle breakdowns in military terms as a constant battle to identify problems and faults before users ride a faulty bicycle and cause further damage to the system.

> We know there are a significant number of bicycles with problems but we do not know the type of problem or where they are; moreover, it is almost certain that these faulty bicycles will continue in circulation, because a bicycle with a problem that does not prevent it from moving will continue to be used on the roads. It is a permanent battle against time and the users, a battle to outpace the deterioration. (Interview with the Head of Sector)

Breakdowns that require an agent to block a bicycle are generally imperceptible to users. Agents must be able to not only manipulate unstable objects but also deal with users who simply want to find a bicycle and reach their destination (Fig. 7.1).

> Sometimes you are repairing a bicycle at a station, and you know that there is another bicycle beside you that you need to repair, but suddenly a user arrives and takes it. But if the client is in a hurry or there are no other bicycles at the station, I'd rather say nothing, if the problem isn't too serious. (Interview with Maintenance Agents)

Breakdowns do not present fixed identifiable properties as per the repair manual, but tend to remain concealed, affecting other parts and with

Fig. 7.1 An agent in the process of looking at his bicycle maintenance schedule while a user is going to take a Vélib' (© Photo: M. Tironi)

other potential effects, on the wheels, the chains or the brakes of other bicycles used by other users. Rather than taking for granted a dualism between human actions (agents who detect failures) and non-human actions (bicycles awaiting classification),[7] agents remain uncertain as to the origin and impact of breakdowns because they do not know when and how these breakdowns may manifest themselves, and who or what is acting. This is what makes the breakdowns a *matter of inquiry*.

> The breakdowns that can be found in the field are infinite, so all we can do is repair them whenever we identify one. But it is one failure out of the thousands that there must be, because the bicycles are not going to tell us 'listen, I have a problem and I am in such and such place' and it is entirely possible that this bicycle circulates for weeks without anyone spotting the problem. (Interview with Maintenance Agents)

Bicycles and breakdowns are referred to as *actors* in the sense that the bicycles can conceal or even reveal their problems. The challenge lies in detecting the breakdowns that are not easily visible and which can be potentially dangerous for users. Agents' work is not limited to repairing breakdowns, but it also consists in anticipating potential anomalies and recognising how these anomalies alter the experience before they become 'major problems' for the maintenance agents:

> Often, it's a case of minor problems, which can be repaired in a few minutes, but the problem is that these can pose a risk in the long term. A bicycle bell may be broken which won't prevent a user from cycling, but it may have consequences in terms of client safety. The thing is that although it's a minor problem and possibly imperceptible to the client, the problem with the bell can become a major problem, just like a flat tyre. So, while it doesn't cause an immediate problem, it can become a problem. (Interview with Maintenance Agents)

Being able to recognise that there is a problem with a bicycle relies on having specialist knowledge and an understanding a normal user would not have. These are problems that are not immediately visible:

> What happens is that lots of bicycles don't pose a problem to users, but there are problems. For example, a wheel may not be securely fixed, but the user doesn't feel any immediate effects, because there are safety mechanisms that minimize the effects. (Interview with Maintenance Agents)

Similarly, this operation of reading traces on the bicycles is clearly demonstrated by the agents, who work in the repair workshops in their ability to recognise the different types of buckling that can affect bicycle frames.

> It is important to not confuse a problem with paintwork or rust with a buckled or broken frame, these are two different things. It's true that one must look very carefully to decide that a frame is broken, because there may be a little warp in the rear wheel, a little bulge or crack. One of the techniques I use to identify a buckled frame is to stand myself in line with the frame and look to see if it's straight or not… So that's the first aspect of broken frame, knowing if it's properly aligned, and to do that I try to find a good position. (Interview with Maintenance Agents)

We see here how the task of reading the traces is a corporal task. The agent needs to handle the bicycle in a specific way to get the right view of the frame. This corporal and sensory knowledge (Henke in this volume; Dant 2010) is not just a way of investigating the traces, it also reveals aspects that would otherwise be impossible to see. Maintenance work therefore depends largely on the ability to identify certain traces (Ginzburg 1989), in order to designate a specific bicycle as being in need of servicing. According to Ginzburg, the trace constitutes a mode of indirect and indicative knowledge of the world: traces do not exist in isolation, but are always a reference to something else. In our case, the traces identified by agents must be interpreted not only in terms of causalities but also of effects and resistances.

We are not suggesting that breakdowns are constructed through observation or classification; or that the breakdowns originate when they are detected by maintenance agents. This type of constructivist perspective ignores what we believe is the most important feature of the breakdowns—their effects. Our aim is to highlight the different ways these breakdowns exist, and, in this case, their state of latency, or state of *plasma*, to use one of Bruno Latour's concepts (2005). Breakdowns are not only present when they are identified or repaired, they are also present in an intermediary state in which they are not totally visible and understandable and, therefore, their effects are not the same. Latour defines the plasma state as what "is not yet formatted, not yet measured, not yet socialized, not yet engaged in metrological chains, and not yet covered, surveyed, mobilized, or subjectified" (Latour 2005: 244).

We are not trying to portray an *essence* of faults whereby breakdowns are static entities readily available for observation. From this perspective, anomalies would be limited to the scheme's technical properties. The failure would be defined by the object's material properties, regardless of how these breakdowns are interpreted. Similarly, we reject the constructivist perspective of breakdowns as social constructions in which breakdowns only exist because they are categorised and anomalies are born from that classification system, whereas in the first perspective it is only the object's essential properties that would deem it 'out of service'. In contrast, we have tried to show that breakdowns are neither confined to either the properties of the object, nor to the classification process carried out by maintenance agents, because, empirically, this distinction becomes blurred in the agents' concrete activities. Maintenance work is not based on static entities but on the temporary traces of breakdowns and bicycles in their various states of transformation and problematisation. To some extent it is this task of making breakdowns 'visible', where human and non-human competences are interwoven, that makes the maintenance of the Vélib' scheme a fragile act of permanent investigation.

Maintenance as Innovation

Examining the maintenance operations revealed that the primary material with which agents must grapple—the breakdowns—are both diffuse and elusive, discreet and progressive, visible and invisible in nature. We could use Dewey's notion (1938) of 'diffuse quality' to refer to the elusive *action* of breakdowns on the Vélib' service. The elusive and non-standardised nature of breakdowns requires agents to adopt an attitude of inquiry and exploration.

Faced with anomalies, agents often find experimental solutions, responses that are reliant on their engagement with action, body and *transactions*.[8] Maintenance capacities are produced through interaction; resources are discovered through the circumstances maintenance agents find themselves in. It is a form of knowledge that cannot be mastered internally—it is not a transferable inventory of information but something that is produced through action. For example, agents utilise the urban environment (station location, time of day, type of people, etc.) in their maintenance work and even to determine the nature of the failures. The following example from my fieldwork is particularly revealing.

Maintenance agent Vincent is responsible for station number 13104, situated at the Quai d'Austerlitz. It is 8:30 am. It is a large station with a capacity of over 60 bicycles. Vincent is working quickly, because he knows that users will arrive soon to take the available bicycles. The agent spends only a few moments answering my questions. I see him concentrating fully on repairing the bicycles and overseeing the station in general. From time to time, he communicates with the head office via his mobile phone to request the blocking of a terminal that he is unable to repair. However, users arrive at the station to collect or return bicycles much faster than he can repair the bicycles. I decide to interrupt the agent to ask him if he can tell what kind of different types of anomalies he will encounter depending on the neighbourhood:

> Generally, depending on the station I am going to work at, I calculate which spare parts to take with me. Depending on the sector, you know more or less which parts will be broken and which will resist. But with maintenance there are no norms, because you can always find different problems, it is impossible to plan ahead and planning and monitoring maintenance is only a guide, because something unexpected always happens and needs to be fixed depending on the situation and each agent does that his own way. (Interview with Maintenance Agent)

Vincent, in some ways, alludes to Dewey's concept of inquiry when talking about his experience with breakdowns. A practice of inquiry inevitably creates links between uncertain or problematic situations and a choice of possible solutions. Therefore, the definition of a problem (or breakdown) is inseparable from the task of inquiry. The problematisation, or identification of a problem, is the first step in the investigation process, "discovering what is or are the problem(s) that a problematic situation poses to the inquiry is to be well-advanced in the inquiry" (Dewey 1938). Hence, Dewey presents the concept of inquiry as an uncertain situation where the thing one is looking for is not known in advance and emerges as the result of a series of different tests. As Stark explains, following on from Dewey:

> There is not at first a situation and a problem, much less just a problem and no situation. There is a troubled, perplexed, trying situation, where the difficulty is, as it were, spread throughout the entire situation, infecting it as a whole. If we knew just what the difficulty was and where it lay, the job of reflection would be much easier than it is… In fact, we know what the problem exactly is simultaneously with finding a way out and getting it resolved. (Stark 2009: 2)

Breakdowns are not inherently a problem, they only become problems when they create an uncertain situation for action. A problem can only be articulated if it is experienced (Dewey 1938). The notion of problem as suggested by Dewey implies an episode of rupture in a subject's activity—a rupture that prompts the search for the necessary means to clarify an uncertain situation and propose a solution to the problem. According to Dewey, a problematic or confused situation raises emotions and questions, putting into motion the actions of inquiry, clarification and elucidation. Without the *experience* of a problem, there is no motivation to create and explore solutions.

For Vincent, the solutions to breakdowns are not determined by a series of pre-set parameters. As he says, 'it is impossible to plan ahead', which takes us back to the premise of an inquiry adapted to a given moment. The truth of a breakdown, its solution, does not precede the situation that may occur. As the agent says, "something unexpected always happens and needs to be fixed depending on the situation". The unstable nature of breakdowns requires field agents to have an open attitude to accept this uncertainty.

This coexistence with uncertainties is also evident in the agents' daily work as a considerable part of their day is spent constantly *adjusting* and experimenting, and, in some cases, actually innovating. This position could equally be described by the term *bricolage*, in the sense of a capacity to generate solutions based on the nature of the breakdown.

One of the most common problems encountered by agents are punctures. Punctures are so frequent that at least once a month, agents need to replace their worn-out bicycle pumps. The seemingly simple task of repairing a puncture is in fact very time-consuming as the bicycle must be dismounted from the terminal and the wheel removed. Vincent explains that these breakdowns are 'laborious' because, while they are technically straightforward, repairing it quickly requires the right position. He describes how through his experience of these types of time-consuming problems, he devised a technique to repair a puncture without removing the wheel. According to the agent, this 'little trick' saves him huge amounts of time.

The technique developed by the agent involves wedging the tyre lever between the edge of the bicycle and the docking terminal (see Fig. 7.2). This trick allows Vincent to elevate the wheel a few millimetres from the ground (see Fig. 7.3), giving enough space to replace the inner tube without having to remove the wheel. Vincent describes the trick as 'a little innovation that saves [him] time and lets [him] fix a typical breakdown that [he] personally finds laborious'.

According to the agent, the routine activities of his job (such as repairing punctures) are to some extent opportunities for innovation.

Fig. 7.2 Agent repairing a puncture (© Photo: M. Tironi)

Fig. 7.3 A tyre lever in the docking terminal (© Photo: M. Tironi)

Breakdowns create a space for exploration, or as Vincent says 'little innovations' in the field. The act of maintenance is closely associated with exploration and in this respect maintenance is learning (Graham and Thrift 2007). Henke (2000) even suggests that improvisation is one of the key elements in the practice of repair (see also Schubert in this volume). Vincent's case also takes us back to the question of the effects of breakdowns on agents' work. Solutions that are not yet formulated open up the possibility of new combinations and associations. David Stark considers that the privileged spaces for innovation are the realms of ambiguity and uncertainty, as this is where actors are compelled to reassemble their knowledge and techniques. To cite Stark: "because innovation, in this view, involves bringing together incompatible traditions, we should not expect that the process will be harmonious" (Stark 2009: 3).

Maintaining and Inquiring About Users

Vélib's materiality (docking terminals, bicycles, etc.) is continually involved in updating the social and spatial order consisting of practices and discourses from users and non-users distributed within the public space. With its exposed nature and city-wide presence, Vélib' interacts powerfully with the urban environment in which users' demands, road traffic regulations, weather conditions, tourist practices and acts of vandalism are interwoven with many other situations that can occur around a station. Vélib' not only occupies a specific urban space, it also shapes it, in the same way that technology is shaped by the urban space in which it is inserted. The identity of the space is not fixed or determined from the outside, but in fact always exists as a possibility constructed from the different objects that equip the city (Law 2000).[9]

Maintenance agents have privileged access to this co-construction between Vélib' and the urban space and actively participate in its elaboration and reproduction. Restricting agents' activities to a material maintenance disconnected from the rest of the Vélib' service would be unfair and empirically incomplete on the part of the analyst. Failure maintenance acts within a continuous process, which needs no distinction between 'material breakdowns and inquiries' and 'user-related failures': the failure is as 'technical' as it is 'social'. There is a close articulation between repair activities oriented towards objects and those oriented towards people. Agents develop specific skills on how to adapt users and technology, and create precise definitions of their practices and difficulties. They must constantly reconcile the projected users (user types) of the Vélib' technology and the real users, for example, the tourists who have never used a self-service bicycle before. Agents not only have an extremely rich sociological definition of users, they also make permanent adjustments in situ to harmonise the technology with the user, and the user with the technology.

Consider the following instance at a station near Opéra, in the 9th arrondissement of Paris: I notice that the agent has started talking to a user who is returning a bicycle to the station (Fig. 7.4). He explains to her that next time she should secure her bag inside the basket as otherwise it is very easy to steal. He shows her how to do this by passing the shoulder straps round the handlebars of the bicycle, explaining that he has seen numerous robberies. The woman is impressed but also grateful for the information given by the agent about how to use the Vélib' bicycles more safely.

Fig. 7.4 Agent talking to a user (© Photo: M. Tironi)

This highlights how agents participate daily in the construction of the Vélib' users. This urban mediation by maintenance agents highlights two important elements. First, the agent reconfigures the relationship between the user and the technology, making the user act differently—because next time our user will probably secure her bag to the basket—and second, the bicycle is no longer just a means of transport, it has also become a *medium* for a potential robbery. Therefore, maintenance agents' work cannot be summarised as simply an 'update' or 'connection' between the conceptual universe and the users' worlds, because through their interventions agents frequently redefine the nature of the scheme.

Thus, the agent prescribes how the bicycle should be used. We know from the works of Akrich (2006)[10] that any innovation project should *position* the technologies within the representations of action and environment in which the technology will evolve. The scripting of future users (*scripts*) by designers and engineers defines an action framework (*framing*), i.e. how 'normal' users should behave with the technology.

Nevertheless, we find that maintenance agents do not simply update the pre-defined scripts (as would be the case of a maintenance agent showing tourists how to rent a bicycle), they also participate in producing new scripts, i.e. the instructions that agents create in situ. The cognitive and corporal process of securing a user's bag to the bicycle's handlebars is not part of the designers' scripts and discourses.[11] It is a process introduced by agents on the basis of what they have learnt in the field. Agents do not limit themselves to updating the scripts (skills, attitudes) that were predefined and projected by the scheme's designers, but are spontaneously creating and defining what needs to be done to become a safer Vélib' user. This appropriation by the Vélib' agents compels them to explore new possibilities and produces developments that enable the programme to be better adapted to its environment.[12] Here we have another form of inquiry into Vélib' users' behaviour: agents are continually determining the best way to maintain/repair certain user behaviours, trying to discipline users to adopt behaviours that the agents themselves consider as safer.

Inquiry into 'Social Breakdowns'

In this section we describe agents' working conditions and schedules and the ways they classify and record 'vandalism'. We see the emergence of a situated knowledge among agents in the way they record vandalism, discuss and contextualise it, and the skills used to understand and deal with it. In this section, we are also interested in looking at the way agents generate and embed narratives and inquiries in relation to acts of vandalism, devising categories about the programme, the users and the urban ecology. We see how these inquiries rely on an extremely local and corporeal knowledge of the different entities that are involved in the functioning of the service.

During our study, agents explicitly expressed their interpretations of the 'antisocial' acts that affect the Vélib' scheme. It is clearly an acute issue for those responsible for the system: records show that by 2010, 8000 bicycles had been stolen and more than 16,000 had been damaged or 'vandalised'. Of the 16,000 bicycles that were vandalised between 2007 and 2010, half had to be replaced.[13]

Although some acts of vandalism are easily recognisable by agents (knife marks in tyres and wheels; bent frames, wheels and baskets), others require a more elaborate knowledge. For example, we are observing a mechanic at the Saint-Denis workshop who is replacing a fork. We step forward to take a closer look, but we do not see anything wrong with the fork. It does not look bent or dented, and we cannot see any problems with the paintwork or any rust. Yet the mechanic is in the process of removing the fork. When we ask him why, he shows us a cut in part of the fork, a cut so thin and clean that it is not immediately visible. We ask him what could have happened to the fork and he simply replies: "the fork has been cut, this usually happens during protests". The fork has a completely perfect cut, as if cut with a scalpel. Unlike other problems, this cut seems almost elegant, or at least uncommonly precise. According to the agent:

> This cut was done intentionally, with a special implement. A cut like this can't be done by just anyone; this was done by somebody who knew what they were doing. We don't know exactly who does this, but it might be people who own bicycle shops. (Interview with Maintenance Agent)

The mechanic's interpretation suggests a conspiratorial act. Cuts with this level of precision cannot be done by "normal users", only by organised individuals with specialist knowledge. Aside from the nature of the interpretation, it is interesting to see how agents construct a story worthy of a detective novel on the basis of a cut in a bicycle fork. Their interpretation conjures up actors and intentions that are not physically present and cannot be proven, but which become traceable through the material evidence and traces that the agent sees on the bicycle. The following statement shows a similar process of investigation, but in this case referring to the social causes of vandalism.

> I have worked here for several years, and for me the issue of vandalism is linked to a question of social differences. Sometimes, I get the impression that we are working to maintain a service for the '*bobos*' of Paris (the 'hipster'). If you think about it, working class people who get up early and go to work at 6:00 or 6:30 in the morning, don't really feel like taking a Vélib', particularly if they work outdoors, often doing manual labour. Also,

people use Vélib' to avoid the morning rush hour in the metro, all the stressed people, which is really unpleasant. But at 6:00 in the morning the trains and metro aren't as busy. It's people who spend all day stuck in their offices who want to use the Vélib'. (Interview with Maintenance Agent)

This raises the question of how Vélib' is perceived by different social groups. According to the agent, vandalism should be considered from this angle. The agent puts a value or principle first and foremost (in this case, social justice relative to different lifestyles) to explain acts of vandalism. He is engaged in a descriptive activity, a system of individual understanding, which will have practical effects on how he organises his maintenance work. Effectively, his understanding of vandalism will act as a situational or empirical anchor to his daily operations. The same agent explained to us:

In affluent neighbourhoods, there is less damage to stations and bicycles than in the working-class neighbourhoods; yet in the working-class neighbourhoods there is more respect and understanding for our work. Users in the centre of Paris are always in a hurry and stressed, they barely say good morning or thank you to us. (Interview with Maintenance Agent)

From his experience in the field, the agent views vandalism in a way we could term empathetic. In working-class neighbourhoods, there is more vandalism, but residents have more respect for agents' work. By explaining damage and vandalism in sociological terms, the agent is attempting to position the Vélib' experience within a broader category. Maintenance agents feel entitled to express their opinions about vandalism, because they feel their experience gives them knowledge of the city and its forms of urbanity.

The type of damage to bicycles varies a lot between the North and South of Paris. In the North you find wilful acts of vandalism, slashed types, bent frames, etc. But there is less vandalism in the city centre because these areas are busier, people don't dare to vandalise, because the ones who break the bicycles are generally non-users. (Interview with Maintenance Agent)

The Notion of Non-user

Let us consider for a moment this reference to the spatial behaviour of vandalism and the notion of the 'non-user'. This term clearly shows the types of situated knowledge agents can produce. The scheme was implemented spatially and socially taking into consideration the users, their preferences and transport needs. Yet, it has become apparent that the vast majority of acts of vandalism have been committed by 'non-users'. This category of users, generated by the maintenance agents themselves, has a significant impact on the way that maintenance work is organised and the fleet of bicycles is managed (Fig. 7.5).

> We know that if we don't repair stations or bicycles for a long time, they will be vandalised, and by people who are not necessarily users but are just passing by, a bit drunk, who mess around with the bicycles, college kids near schools, etc. (Interview with Maintenance Agent)

Fig. 7.5 High school students at a Vélib' station 9th district in Paris (© Photo: M. Tironi)

We do not know if the 'non-users' category refers to people who feel excluded from the system for technical reasons, or people who resist using it for social or other reasons. Nevertheless, this illustrate how agents have devised this category based on their experiences in the field. The act of categorising a certain profile with specific characteristics and moral intentions requires an ability to make links and associations, generate scenarios of behaviours that are shared by certain groups. However, this 'construction' of profiles is not done during the service's design phase by experts, spokespeople or through statistical studies. Agents construct this category of users in action, based on the social, spatial and material *inquiries* available in the field.

Vélib' Faced with Mass Crowds

Strikes and mass protests create testing situations for the entire service. Faced with emotions running high and crowds of people who are not necessarily Vélib' users, the service experiences a technical and social transformation. During our study, national protests against the government's pension reforms turned Paris into a stage for numerous public demonstrations. Maintenance agents know that however much they may plan and prepare prior to a demonstration, it is never enough to deal with the aftermath of an event on this scale. Strike days are good for Vélib' in terms of the number of journeys made (as other modes of transport are limited or out of service), but they also result in the highest number of acts of vandalism.

When the strike ends, the agents reopen all the stations that were closed for security reasons and restore order to the Vélib' network. As the city tries to return to normality, the aftermath of the crowds of protesters becomes apparent. Consider the following fieldnotes:

> At about 3:00hrs I join one of the maintenance agents responsible for opening the stations in the Bastille area. Everything seems normal and apart from some flyers and stickers on the ground, there is no indication that there was a large protest here the night before. Just then the agent receives a call on his mobile from his Head of Operations (HO).

HO: What's the situation? Much damage?
MA: I'm just removing the orange plaques from station number…and then I'm going to check the electronic problem at the other station.
HO: Ok, and not too much damage?
MA: Normal, not much in the stations at the moment but lots of wrecked bicycles, all over the place. I've found three at least already.
HO: Fine, let me know when you fix the problem at the station.

This conversation reveals the mood of Vélib's senior management after a public protest. Knowing that after a rally there will be an unusual amount of damage, this is the first question the Head of Operations asks the agent. But what does the agent mean when he talks about wrecked bicycles? Bicycles vandalised by protesters? Agents will interpret these acts of vandalism as the consequence of emotions running high.

> On strike days there is more demand, and of course more people use Vélib' when there's no public transport, so we have to constantly replenish and free-up space in the stations. But it's a very precarious time because people are agitated so if they don't find a space to leave the bicycle, they may become violent or just leave the bikes anywhere. That's why when there are protests you find damaged bicycles all over the place. (Interview with Maintenance Agent)

It is not always 'non-users' who cause damage, but, as the agent says, users of the service who are in an agitated state. Given that public transport runs less frequently on strike days, demand for Vélib' increases significantly, as does the desperation of users trying to find or return a bicycle. This is when the diffidence and civic behaviour of the Vélib' user gives way to emotional outbursts; urbanity and respect are replaced by small acts of disobedience (Fig. 7.6).

> Some users will make long journeys with a bicycle that is in a bad state, I've seen this on strike days and I really don't understand it, people lose all respect for the service, and for me it's almost as bad as breaking a bicycle. (Interview with Maintenance Agent)

Fig. 7.6 Bicycles with a puncture and bent frame (© Photo: M. Tironi)

Although the act of continuing to use a bicycle even though it has a flat tyre may seem harmless, for the agents it is an antisocial act that causes as much damage as someone willfully damaging a bicycle. The urbanity of the user seems to be temporarily suspended as emotion and agitation invade the public space. Maintenance agents witness this change amongst the users and also from the state of the bicycles left abandoned in the streets. What is interesting about the agents' descriptions is the way they describe the *work of emotions*[14] on the Vélib' during protests.[15]

When De Laet and Mol (2000) studied the case of the water pump introduced in Zimbabwe, they noted that the success of a technology lay in its capacity to adapt to a situation, to make its properties malleable according to different environments. The authors suggest that the pump's adaptability is in its capacity to constantly renegotiate its limits (material, hydraulic, social and geographic), and in the integration of different elements of its environment into its operation (De Laet and Mol 2000: 252). From this perspective, the acts of vandalism endured by the Vélib' scheme are part of the numerous socio-technical interactions between technology and its environment. Vélib's permanence depends on its capacity to resist and mutate according to the event or situation it faces. Its stability is not based on a 'black box', but depends on its capacity to adapt to an adverse and unstable environment. The infrastructure must be inserted in the city in a way that is sufficiently robust to transport thousands of clients every day but also sufficiently flexible to be able to modify the operational protocols that ensure its permanence. Within this process, maintenance agents' inquiry work is part of the laborious adaptation process that Vélib' faces day after day. In their activities, agents produce knowledge that allows them to evaluate the condition, breakdowns and transformations of the Vélib' programme.

Conclusion: Urban Ecology as a Place for Experimentation and Inquiry

By emphasising the significance of the maintenance and repairs agents' operations and categorisations, we are able to understand the Vélib' city not as a stable and intrinsic entity, but as an entity that must be continuously created. The city, as a fragile event, invites us to move from a logic of beings to a logic of becoming, recognising the complexity, heterogeneity and the unthought of urban space. The aim is to trace the heterogeneous trajectory of Vélib' technology in the act of maintenance rather than through seemingly stable and ready-to-use objects. Through the maintenance operations we have attempted to examine one of the Vélib's modes of existence, showing that these distributed and local practices influence the durability and reproduction of the system. As Denis and Pontille (2010) suggest, this approach allows us to problematise the question of modes of existence and the development of objects, as "maintenance takes effect when people develop the capacity to perceive differences within a system of objects that may appear stable at first glance". Denis and Pontille show that the static relationship (in terms of immobile objects) between users and the Paris subway signs changes when viewed from the perspective of those responsible for repairing the signs, for whom these objects are subject to permanent wear and tear and transformation. In this paper, we have observed something different: whereas the relationship between the majority of Vélib' users and the bicycles is one of *mobility* (objects which they expect to find ready for a journey), we discovered that the maintenance agents relate more to the immobility of the objects, as it is only when the bicycles are stationary that agents can carry out their repair work.

To some extent this chapter is in keeping with the sociological approaches to technology, which have extensively described the way in which designers and producers think about the future development of a technological object.[16] These studies have revealed how technological objects have an extremely delicate implicit 'sociology' in regard to the behavior of clients and users of future technologies. They tend to focus on the role of technology innovators and designers, the controversies engendered by these technologies, and the negotiation in the design process; however, they sometimes give the impression that the implementation of these technologies is the prolongation of a network built by the designers of the innovation (Vinck 1995).

We have shown that the life of an object is reconfigured during the field deployment phase, which poses, once again, the question of its identity and its biographical possibilities in terms of its articulation with the environment and contact with users (Akrich 2006). Through their work, maintenance agents are continuously re-specifying the predetermined scripts (for example, the function of the bicycle basket) redefining the frontiers between the design world and the deployment world. As Vinck (1995) and Bovet and Strebel (Chapter 4 in this volume) note, the constitution of an artefact or infrastructure is not only the result of design and manufacturing activities, but extends through the phases of deployment, use, repairs, and maintenance, as it is the latter two of these operations that make it possible to recontextualise and give *livability* (Bovet and Strebel, Chapter 4) to objects, acknowledging the ways in which objects are socialised and used, given meaning and adapted to the urban environment.

Therefore, instead of talking about 'interpretative flexibility',[17] where the Vélib' object would remain stable amidst various superimposed interpretations, we should focus on the idea of *multiple ontology* (Mol and Law 1994) which allows us to consider the mutations of the object consistent with the practices in which it is inserted. Vélib' acquires a distinctive *socio-material identity* in its maintenance and repair practices which is primarily characterised by breakdowns and anomalies rather than by stability and movement. This leads us away from a pre-existing and perfectly demarcated definition of the Vélib' object, in order to observe the way in which different operations make Vélib' exist and act differently.

Various studies deal with the question of *innovation intermediaries* (Von Hippel 2005; Howells 2006; Stewart 2007), but leave open the question of the potential role of *maintenance agents* in reconfiguring and adapting technology to its environment. These studies reveal the presence of various intermediaries in an innovation process (users, organisations, technologies, distributers, buyers, advertising agencies, consultants, etc.) whose role in the adoption of the innovation can be seen, for example, in activities of knowledge production, inspections or dissemination. With the exception of the concept of lead user developed by Von Hippel (2005) these studies give preference to intermediaries with a 'technical background', i.e. experts who do not play a 'supporting' role like that of the maintenance agents in this study.[18] Therefore, in answering the key question of the asymmetry between users and designers,[19] or the alignment between technologists and users, responses are based on feedback provided by users or other comparatively invisible intermediaries, but not on the maintenance and repair practices.

In this chapter, the argument has been that it is possible to extend the notion of *innovation intermediaries* by including the maintenance and repair agents' activities, showing how these operations can be conceptualised as *forms of inquiry* that respecify the infrastructure's durability. These agents must be able to articulate, in their maintenance and repair work, the designers' or programmers' original representations of users, while also exploring new ways of furnishing these concepts based on actual events and their field observations. Through their inquiry in the city and with the scheme, agents become specialists on users and actively participate in defining what a user of a public bicycle system is and should be. Maintenance agents have to adapt the notion of Vélib's 'projected users' to the 'real users', adjusting the scheme's technical choices to its real, everyday environment. In their repair operations, maintenance agents update not only their knowledge of the scheme, but also their knowledge of the multiple interactions between the scheme and its environment. In this respect, our aim has been to show that maintenance work is inseparable from the relatively formal inquiry operations. Maintenance and repair is not simply an action that seeks to restore the original status quo; repair is in fact an operation that permanently re-interrogates the scheme's materiality and its multiple interactions with its environment, producing new knowledge and hypotheses in relation to the service's socio-technical ecology.

Therefore, we say that the agents act as mediators, in the sense Hennion (2002) attributed to this term: the agents are not simply carriers of standardised information about the field, or intermediaries between designers and users—they develop new information and associations for the scheme through their maintenance work, proposing new definitions of action and users, the urban environment and its interconnections. This allows us to reconsider the trio of designer-intermediary-user not as ontologically stable figures but as states or trajectories that are revealed through the tests carried out on the Vélib' object. There are not two distinct worlds: a world of ideas and inspiration exclusive to the design phase and another that we could call the deployment world that is limited to replicating pre-established ideas. The maintenance agents show us that this contradiction between ideas and their execution is far more complex and is subtly dispersed within a permanent cycle.

For the agents, the urban ecology (with its stations, spaces, failures, bicycles and users) is a place for exploring and experimenting new adjustments, a space populated with uncertainty and frictions, which agents must repair and maintain. In his works on the city of Chicago (1929),[20] Robert Park established the notion of the city as a place of experimentation, where the researcher tries to intervene in the urban ecology from a material and normative perspective. Inspired by pragmatism, he proposed the idea of the city as a 'social laboratory',[21] describing the different forms of urban life from this perspective. Thus, the city is a privileged place for a social science researcher to experiment by analysing the protocols and instruments that construct and maintain the urban territory. If we take the role of maintenance agents and the different inquiries they use in their field work seriously, it is because, like the social science researcher described by Park or the scientist evoked in the 'laboratory studies', the agents are constantly redesigning the properties that give life to the Vélib' experience—collecting data and testing the state of the bicycles, upgrading stations and guiding users, maintaining wheels and chains. By using their knowledge, instruments, inquiries and representations of users, agents play a key role in the composition and opening of the Vélib' technology. Vélib's reality is experimental because it is only through the agents' experience and intervention in the field that the infrastructure will find temporary forms of stability.

Therefore, instead of considering the notion of 'urban laboratory' as a rigid label for classifying a 'smart' initiatives, we have tried to convey it on the scale of the ordinaries and mundane practices of maintenance. The proliferation of the inquiries made by maintenance agents to tackle breakdowns and frictions, reveals a type of "laboratory" (Park 1929) in which the urban condition is not only made of solutions, but of problems and failures. This perspective requires us to reject the notion of experimentation as a position of principle that is external to the actors, and rather understand experimentation as a way to deal with the complexity and excesses of urban life (Marres 2018). At the same time, this approach invites us to study the infrastructures of the so-called smart city from the perspective of its varying degrees of temporary and material opening, observing its successive reconfigurations in the field. In sum, exploring maintenance and repair practices can be conceived as a way of re-thinking the multiple modes of being involved with and making the city. Paying attention to maintenance activities is as mode of exploring new narratives for the city, incorporating care and experimentation as central elements, allowing the frictions of the city to (in)form new possibilities on the composition of the urban life.

Acknowledgements Martin Tironi's involvement in the writing of this paper has been supported by the Chilean National Fund for Scientific and Technological Development's (FONDECYT NO. 1180062). The author wishes to thank Ignaz Strebel, Alain Bovet and Philippe Sormani for their insightful comments on the different versions of this paper.

Notes

1. It is important to note that Vélib', launched on 15 July 2007 and managed by the French advertising corporation JCDecaux, was replaced by another concession on December 2017, called Vélib' Métropole.
2. It is important to note that the resources invested in opening a black box are always proportional to the resources used to close it (Latour 1987).

3. The rise of this perspective (mobility turn) has resulted in the proliferation of a series of terms, such as society-network (Castells 1999), liquid modernity (Bauman 2000), mobility environments (Bertolini and Dijst 2003) homo mobile (Amar 2010), nomadic metaphysics (Cresswell 2006), among others.
4. Citing Harman's work (2002), Graham and Thrift (2007) indicate that immersion in places of disconnection and dysfunction is only possible if the idea of connection and coherence is acknowledged.
5. With the exception of Figs. 7.1 and 7.2, all images were taken during our field observations.
6. To date there have been 16 meetings of the Vélib' Users Committee which have dealt with different matters put forward by the 20 committee members. There have also been guided tours to educate the Committee about the internal workings of Vélib'.
7. For a theoretical development on the question of dualism, see notably Callon (1986) and Latour (1994).
8. Again we borrow Dewey's concept of 'transaction' (1938) by emphasising a specific feature of the maintenance agents' activities, which is their insertion and continuity within the city's urban ecology. Dewey perceived experience as a permanent transaction between an organism and its environment, superseding the dualism that tends to consider these two entities as autonomous and independent, and therefore, posing the idea of the *continuity* or the *phases* of the experience.
9. This critique of the absolutist conception of space is found in the works of Nigel Thrift (1996) and the article by A. Mol and J. Law (1994).
10. See also Wilkie and Michael (2009).
11. In the Vélib' user's guide, which can be found at http://www.velib.paris.fr/Comment-ca-marche/La-securite, the only reference to theft stipulates the following: 'when stopping, even for a short time, attach the bicycle to a parking hoop with the padlock provided'. No reference is made regarding how users should look after their belongings in the basket.
12. Julian Orr (1996) developed a similar argument, by showing that the repair agents do not necessarily follow formal manuals and procedures provided by the hierarchy, and that their knowledge of repairs is very often gained and improved through coffee break conversations.
13. In May 2010, the Paris Mayor's Office launched a public awareness campaign on 'user care' to tackle vandalism, with the slogan 'Beating up a Vélib' is easy…it can't defend itself'.

14. Louis Quéré (2012) uses the term 'work of emotions' to refer to the role the emotional dimension can play in the formulation, perception, definition and solution of a public problem.
15. The emotions associated with Vélib' are part of the multiple activities created by the technology. In the song by Philippe Katerine called *Parivélib'*, it says: 'If I can give you a piece of advice, take the Vélib' at night on ecstasy'. Agents also describe night time as a particularly complicated time due to inebriated users.
16. Here we are primarily referring to the works of the SCOT tradition (Social Construction Of Technology), initially led by Wiebe Bijker and Trevor Pinch (1987) and the development of the 'Safety Bicycle'. For a complete analysis of these approaches, see Vinck (1995) and MacKenzie and Wajcman (1999).
17. This expression refers to the different interpretations attributed to a technological object during the concept and design phase. The success of an artefact is the result of the reduction of this process of 'interpretative flexibility' (Pinch and Bijker 1987).
18. Orlikowski (2000), in an attempt to move beyond the linear scheme of 'producing' and 'consuming', analyses the role of technology 'installers' (or support staff) by examining various empirical cases within organisations, however these installers are still typically expert consultants.
19. Boullier (1997: 13) clearly suggests: "the dominant characteristic of users supports the asynchrony of adjustments between the 'end user' and the 'designer-producer'."
20. For an analysis of the Chicago School, see Gieryn (2006).
21. See Tironi and Laurent (2015).

References

Akrich, Madeleine. 2006. Les utilisateurs, acteurs de l'innovation. In *Sociologie de la traduction. Textes fondateurs*, ed. Madeleine Akrich, Michel Callon, and Bruno Latour, 253–266. Mines: Presses des Mines.

Amar, Geroges. 2010. *Homo Mobilis. Le nouvel âge de la mobilité. Éloge de la reliance*. Paris: Éditions Fyp.

Bauman, Zygmunt. 2000. *Liquid Modernity*. Cambridge, MA: Polity Press.

Bertolini, Luca, and Martin Dijst. 2003. Mobility environments and network cities. *Journal of Urban Design* 8 (1): 27–43.

Boullier, Dominique. 1997, June. *Les usages comme ajustements: services propriétaires, moteurs de recherche et agents intelligents sur Internet*. Colloque: Penser les usages, Bordeaux-Arcachon.

Callon, Michel. 1986. Éléments pour une sociologie de la traduction. La domestication des coquilles Saint-Jacques dans la Baie de Saint-Brieuc. *L'Année sociologique* 36: 169–208.

Castells, M. 1999. *La era de la información: economía, sociedad y cultura*, vol. I: La sociedad red. Madrid: Alianza Editorial.

Cresswell, Tim. 2006. *On the Move: Mobility in the Modern Western World*. New York: Routledge.

Dant, Tim. 2010. The work of repair: Gesture, emotion, and sensual knowledge. *Sociological Research Online* 15 (3): 1–7.

De Laet, Marianne, and Annemarie Mol. 2000. The Zimbabwe bush pump: Mechanics of a fluid technology. *Social Studies of Science* 30 (2): 161–185.

Denis, Jerôme, and Denis Pontille. 2010. *Petite sociologie de la signalétique. Les coulisses des panneaux du métro*. Paris: Presses des MINES.

Dewey, John. 1991 [1938]. Logic: The theory of inquiry (reprinted). In *John Dewey, the Later Works, 1925–1953, Volume 12: 1938*, ed. Jo Ann Boydston. Carbondale and Edwardsville: Southern Illinois University Press.

Florida, Richard. 2002. *The Rise of the Creative Class: And How It's Transforming Work, Leisure, Community and Everyday Life*. New York: Basic Books.

Gieryn, Thomas. 2006. City as truth-spot: Laboratories and field-sites in urban studies. *Social Studies of Science* 36 (1): 5–38.

Ginzburg, Carlo. 1989. Indicios. Raíces de un paradigma de inferencias indiciales. In *Mitos, emblemas e indicios. Morfología e historia*, 138–175. Barcelona: Editorial Gedisa.

Graham, Stephen, and Nigel Thrift. 2007. Out of order: Understanding repair and maintenance. *Theory, Culture and Society* 24 (3): 1–25.

Grosjean, Michèle, and Jean-Paul Thibaud. 2001. *L'espace urbain en méthodes*. Marseille: Editions Parenthèses.

Hannam, Kevin, Mimi Sheller, and John Urry. 2006. Editorial: Mobilities, immobilities and moorings. *Mobilities* 1 (1): 1–22.

Harman, Graham. 2002. *Tool-being: Heidegger and the Metaphysics of Objects*. Chicago, IL: Open Court.

Henke, Christopher R. 2000. The mechanics of workplace order: Toward a sociology of repair. *Berkeley Journal of Sociology* 44: 55–81.

Hennion, Antoine. 2002. *La pasión musical.* Buenos Aires: Ediciones Paidós.
Howells, Jeremy. 2006. Intermediation and the role of intermediaries in innovation. *Research Policy* 35 (5): 715–728.
Latour, Bruno. 1987. *Science in Action.* Cambridge, MA: Harvard University Press.
———. 1994. On technical mediation—Philosophy, sociology, genealogy. *Common Knowledge* 3 (2): 29–64.
———. 2005. *Reassembling the Social.* London: Oxford.
———. 2009. Les moteurs immobiles de la mobilité. In *De l'histoires des transports à l'histoire de la mobilité*, ed. Mathieu Flonneau and Vincent Guigueno, 7–10. Rennes: Rennes: Presse Universitaire de Rennes.
Latour, Bruno, and Émilie Hermant. 1998. *Paris Ville Invisible.* Paris: La Découverte.
Law, John. 2000. *Objects, Spaces and Others.* London: The Centre for Science Studies, Lancaster University.
MacKenzie, Donald, and Judy Wajcman (eds.). 1999. *The Social Shaping of Technology.* Buckingham: Open University Press.
Marres, Noortje. 2018. What if nothing happens? Street trials of intelligent cars as experiments in participation. In *TechnoScience in Society, Sociology of Knowledge Yearbook*, ed. Sabine Maassen, Sascha Dickel, and Schneider. Nijmegen: Springer and Kluwer.
Mol, Annemarie, and John Law. 1994. Regions, networks and fluids: Anaemia and social topology. *Social Studies of Science* 24 (4): 641–671.
Orlikowski, Wanda. 2000. Using technology and constituting structures: A practice lens for studying technology in organizations. *Organization Science* 11 (4): 404–428.
Orr, Julian. 1996. *Talking About Machines: An Ethnography of a Modern Job.* Ithaca, NY: Cornell University Press.
Park, R. 1929. The city as social laboratory. In *Chicago: An Experiment in Social Science Research*, ed. T.V. Smith and L.D. White. University of Chicago Press.
Pinch, Trevor, and Wiebe Bijker. 1987. The social construction of bakelite: Toward a theory of invention. In *The Social Construction of Technological Systems: New Directions in the Sociology and History of Technology*, ed. Wiebe Bijker, Thomas Hughes, and Trevor Pinch, 159–190. Cambridge: MIT Press.
Quéré, L. 2012. Le travail des émotions dans l'expérience publique: marées vertes en Bretagne. *Raisons pratiques*, 135–162.

Star, Susan Leigh. 1999. The ethnography of infrastructure. *American Behavioral Scientist* 43 (3): 377–391.
Stark, David. 2009. *The Sense of Dissonance: Accounts of Worth in Economic Life*. Princeton, NJ: Princeton University Press.
Stewart, James. 2007. Local experts in the domestication of ICTs. *Information Communication and Society* 10 (4): 547–569.
Thrift, Nigel. 1996. *Spatial Formations*. London: Sage.
Tironi, Martin, and Brice Laurent. 2015. A field test and its displacements: Accounting for an experimental mode of industry innovation. *Journal of Co-Design* 11 (3–4): 208–221.
Tironi, Martin, and Matias Valderrama. 2018. Unpacking a citizen self-tracking device: Smartness and idiocy in the accumulation of cycling mobility data. *Environment and Planning D: Society and Space* 36 (2): 294–312.
Vinck, Dominique. 1995. *Sociologie des sciences*. Paris: Armand Colin.
Von Hippel, Eric. 2005. *Democratizing Innovation*. Cambridge: MIT Press.
Wilkie, Alex, and Mike Michael. 2009. Expectation and mobilisation enacting future users. *Science, Technology and Human Values* 34 (4): 502–522.

8

Interruptions, Lunch Talks, and Support Circles: An Ethnography of Collective Repair in Steam Locomotive Restoration

Christophe Lejeune

Introduction

Participative, collaborative, and social Web advocates are celebrating the advent of self-organized online communities of amateurs producing valuable content or artifacts (Lejeune 2011a). However, in many different contexts, enthusiasts have been coordinating collective efforts long before information technologies offered to support it. Indeed, among others, volunteers have been preserving steam locomotives for half a century. Heritage railways depend on a constantly accommodating group of volunteers, who immanently combine their activities. This paper addresses an intrinsically collective phenomenon.

Regarding repair and maintenance, heritage railways offer singular settings. Unlike museums dedicated to industrial legacy, they do more than preserving decommissioned steam engines. In addition, they

C. Lejeune (✉)
University of Liège, Liège, Belgium
e-mail: christophe.lejeune@uliege.be

I. Strebel et al. (eds.), *Repair Work Ethnographies*,
https://doi.org/10.1007/978-981-13-2110-8_8

intend to operate steam locomotives and run antique trains. Therefore, they fall under conflicting constraints bound to preserving while using.

This chapter sets out to shed light on heritage railways' "invisible" activities. As I will show, although very efficient, the social organization of these activities is widely informal and, as a consequence, go unnoticed at first glance. Of course, this first invisibility is common to lots of, if not all, social activities. Secondly, the people involved in preserving steam locomotives are not trained professionals, but volunteers. Their non-profit organization is neither a craft corporation, nor a national railway company, nor a large-scale private industry. They do not benefit from the prestige of traditional crafts, public service values, or corporate image. This study is, thus, also an opportunity to underline what they do. Finally, focusing on repair and maintenance, this ethnography covers background activities. Service operators are, by design, concealed to the public (Goffman 1956: 71; Thibault 2013: 23–24, 92, 119). While narratives are subject to retrospective rationalization, ethnography permits to open the black box of these background activities. Shadowing members, I gained access to the actual gestures, discussions, and practices as they unfold behind the walls of a messy workshop.

Dissecting four ethnographic vignettes, I will exhibit different ways to organize, regulate, and adjust repair and maintenance through interruptions, lunch talks and support circles. In the meantime, the creativity demonstrated by members facing unexpected difficulties will question the apparent paradox of the continued use of heritage engines thanks to modern technologies.

A Heritage Railway

Across Europe, volunteers are working together to preserve and restore railway lines and rolling stock, focusing mainly on steam locomotives, as well as railcars, coaches, and wagons. The *Three Valleys Steam Railway* is one such non-profit organization, and one of the most important in Belgium. The organization owns a collection of rolling stocks, which has been decommissioned by the national railway companies, such as *Société Nationale de Chemin de Fer* (SNCF in France or SNCB in

Belgium). The organization also rents an abandoned rail line from the national railway company. Located in the Belgian countryside, close to the South-West border with France, the line is 14 kilometers long and serves three intermediary stations. At the beginning of the railroad, the first station has a preserved roundhouse where steam locomotives are sheltered. Coaches, wagons, diesel shunters, and railcars can be found on the tracks around the roundhouse. A welcome desk has been built close to the roundhouse. At the other end of the line, a train museum, a cafeteria, and a large workshop surround the final station. A collection of locomotives stand in the yard around the shop.

During the summer, this rolling stock is driven along the railway line. Excursions are organized for tourists. This generates the income needed to fund the association's restoration projects and enables it to maintain a high profile. Throughout the year, on weekdays, the workshop houses a Work Training Center (*Centre de Formation par le Travail*). This center allows unemployed people to acquire manual know-how with the purpose of applying these new skills to jobs in the industry. On Saturdays, the shop is devoted to the restoration of rolling stock by members of the organization. Although it employs a few paid workers, most of the people working for the organization are volunteers, the majority of whom do not have a professional background in the railway sector. Of course, retired railwaymen also take part in the activities of the association. Some members are still employed by the railway companies, but they are not in the majority. Of the 100 or so members, a variety of profiles can be seen. Although most of the members are men, a few women are active in the organization. Most of them perform gendered activities, such as preparing food, serving drinks, selling tickets, and welcoming visitors. However, a few women also run heavy engines, such as steam locomotives.

As it is often the case in the voluntary sector, a large number of retired people are involved. A wide range of manual professions is represented, including industrial workers, lorry drivers, technicians, gardeners, and tradesmen. The organization's ranks also include retailers and freelance workers (electronics consultants), those from intellectual professions (teachers, service sector managers), craftswomen (blacksmiths), and from the artistic professions (painters, orchestra conductors).

The activities of the organization are relatively varied, as they include coordinating a large number of people, maintaining and repairing rolling stock (in the workshop), looking after the infrastructure (buildings, bridges, tracks, and points), getting the locomotives and trains moving (preparing, coupling, driving, dispatching), contacting sponsors, organizing and promoting the events, welcoming visitors, selling tickets, and accompanying trains.

Fieldwork

Operating a decommissioned locomotive on an abandoned railroad requires a team. A driver and a fireman, occasionally assisted by a supplementary crewmember, run the locomotive while one or two conductors accompany passengers in the coaches. However, this front-line crew would not succeed without a large team of volunteers working behind the scenes. This invisible team carries out maintenance and repairs, either in the shop (primarily where the engines are concerned) or on the railway itself (tracks and points require maintenance, too). This chapter's ethnographic vignettes focus on engine repairs (a diesel railcar and a steam locomotive). These do not cover the large range of activities required to operate the train.[1] However, even restricted to this scope, activities are far from scarce.

I have observed this association's restoration work first hand. I have taken part in its Saturday activities once a month for three years. My frequent attendance makes me a regular, though not its hardest working member (the most committed volunteers are generally present every Saturday). My early contact with the organization, one year before the fieldwork began, suggested that members did not welcome rail enthusiasts exclusively interested in operating steam locomotives. Instead, commitment to repair and maintenance work in the shop was requested. To prove that I was not going to waste their time, I decided to conduct my observation mainly in the shop. This decision surely helps me to be accepted as a peer. It also determined the kind of repairs I witnessed.

I excluded myself from the crews operating the locomotives and performing quick and dirty fixes in-line. Moreover, living far from the shop (four hours by train), I rarely attended early Saturday meetings where activities in the shop were planned and organized. Quick and dirty fixes, on the one hand, and planning meetings, on the other hand, were thus excluded from my fieldwork. The present chapter therefore does not address these phenomena. Of course, once I became aware of the consequences of my early decisions, I reorganized my fieldwork: in 2015, I started training to become a member of the crew that runs steam locomotives (and thus perform in-line quick and dirty fixes). At the same time, I took lessons to obtain my driving license that allows me to drive and attend the Saturday morning organizational meetings. These two decisions, though, concern my subsequent (currently ongoing) fieldwork, that this chapter does not cover. Instead, I focus on elements from the first stage of fieldwork, i.e., repair-in-action.

This study is based on empirical material presented through four ethnographic vignettes. The first two vignettes concern a railcar, while the second two relate to a steam locomotive. I was present and participated in these restorations.

Straightening the Brake Linkage

First vignette

> On 27 April 2014, I was on my way to the heritage railway. As always, I was excited. But that day was special. A few days before, I made a phone call to Oliver, a member of the organization. He told me that, two weeks before, on 13 April 2014, something very unusual, though not too serious, happened to his favourite railcar: it was derailed. No one was injured, fortunately, and hardly any material damage was reported.
>
> Oliver was driving the railcar when the incident occurred. He was surprised and wondered what could have caused this derailment. The incident happened in a place that is not particularly difficult and the train was traveling relatively slowly. Oliver therefore decided to examine the entire railcar, in order to identify the reasons that caused its derailment. He invited me to participate in the diagnostic session, which involved performing a major overhaul of the concerned engine. Such an overhaul differed from the setting I typically observed and participated in the association's shop.

This vignette concerns a French diesel railcar resembling a long bus on rails. I first came into contact with this railcar when I began my observations, on the first day spent at the workshop, on 8 September 2012. As a novice, I was asked to remove paint marks from the windows of the railcar that I will discuss here.

During the phone call, Oliver gave me an appointment at the initial station (while other volunteers typically gather at the shop adjacent to the final station). The railcar was positioned over a pit of the same type as those found in the workshop, apart from one detail – this pit is outside and therefore not located in the actual workshop. The two association members that I accompanied, Oliver and Ian, started by looking at the brake linkage. The linkage is a worm screw that holds the brake shoes and wheels in place.

The end of this threaded rod was slightly bent, probably because it came into contact with the rail when the railcar was derailed. The first repair would therefore involve the straightening of this metal rod. The method used was very instructive. One of the two members picked up an old tube that was lying around inside the locomotive shed. It was actually a steam locomotive tube, through which hot gases from the firebox would pass, before reaching the smoke box (the functioning of a steam locomotive will be described in detail below). This tube was used as a tool, in an attempt to straighten the brake linkage. Sliding the end of the brake linkage into one end of the tube meant that the volunteer had enough leverage to straighten the linkage with his bare hands.

However, this first attempt failed. In fact, as the linkage was relatively close to the bottom of the railcar body, the tube touched the bottom of the body and there was not enough space to straighten the threaded rod.

The volunteers stared at each other. "We need a car jack," they said in unison. Ian asked me to get a few pieces of wood from a junk pile next to the roundhouse. Meanwhile, he went to look for a jack in his own car, which would normally be used to change a tire. As Fig. 8.1 shows, he positioned the jack on the rail thanks to a few pieces of wood and placed a small one between the top of the jack and the end of the threaded rod. Doing so, he grumbled, "I told them a dozen times that the shed should be provided with an industrial pneumatic jack." He then turned the crank on the jack, in order to lift its end, which gradually straightened the threaded rod. While Ian was about to release the pressure on the crank, the two members then noticed the threaded rod was returning to its

Fig. 8.1 Operating the car jack (© Photo: C. Lejeune)

initial bent position. They instantly diagnosed a "spring effect." Without a word, Ian made an additional turn of jack handle to counter this tendency. The threaded rod had now returned to its rectilinear position in alignment with the entire linkage.

Improvizing Tools

This vignette illustrates a typical type of repair. There had been an accident and breakdown (the railcar was derailed), which led to parts being altered and adversely affected (the brake linkage was bent). Because steel is quite flexible, the rod did not break during the derailment. The same reason rendered its straightening possible. In this case, nothing was irreparably broken; instead, the bending was reversible with nothing more than a simple tool and some manpower. With the aid of a car jack, the railwaymen straightened the bent rod to its original position.

The work completed by the volunteers involved restoring the threaded rod, which meant returning it to how it was before the problem arose. This vignette resembles the work completed by technicians in a car garage (Dant 2005: 109). In both cases, the task consists of returning an object to a previous state. Such restorations match the most straightforward definition of repair.

This restoration required no additional material other than manpower and tools. Nevertheless, unlike a car garage or the shop (located at the opposite end of the line), the roundhouse is not provided with dedicated tools. The volunteers must improvise—a car jack replaces an industrial jack, and a boiler tube and some pieces of wood are recycled from scrap material. This interesting phenomenon surrounding the repair and maintenance work is similar to what is documented by the editor in chief of *Make* magazine and advocate of the Do-It-Yourself (DIY) philosophy. While building a chicken coop, Mark Frauenfelder (2011: 99) scavenges his junk pile for needed materials. He discovers the immense resources that can be extracted from scrap material. Indeed, tinkering rests on odds and ends, as stated by Claude Lévi-Strauss (1962: 36). Keeping, gathering, and collecting miscellaneous items allows for the self-reliant creation of a substitute when tools or spare parts are lacking.

In a different context, the required manpower could have been provided by a single person. For instance, Mark Frauenfelder insists that he built his chicken coop on his own. Similarly, some car technicians perform similar repair and maintenance alone (Dant 2005: 111 & sq.). The setting is, however, clearly different, as a locomotive covers a larger space, thus allowing more than one person to gather in the pit, for instance. But material dimensions, such as size, are not sufficient to explain why locomotive restorations are collectively achieved. Unlike technicians, most volunteers are not trained in mechanics. Working in teams allows for the knowledge, skills, and know-how to be shared, spread, and transmitted among volunteers. The bent rod is restored by a team of two close members, assisted by an apprentice—me. For two old friends who have known each other for 38 years, little conversation is required to decide what to do: substituting a car jack for a pneumatic jack, straightening more than was apparently needed to counter the spring effect, or using a boiler tube as a lever. Sharing comparable experiences, they cooperate smoothly. Resting on ingrained habits, their longstanding complicity allows for a fine-tuned mutual adjustment and prevents further issues.

Finally, the two co-workers performing this restoration appear autonomous, being efficient while resting merely on their shared experiences. Through a similar issue, the next vignette shows that they are less isolated than they first appear.

Tightening the Bogie Cables

Second vignette

"Wait a minute," said Oliver while leaving Ian and myself for a moment. Back from his car, he showed us a complete original technical documentation for the railcar. During the two weeks following the derailment, Oliver sought out the origin of the problem. Other associations that owned a former railcar of this type, together with his contacts at the French national railway company (SNCF), enabled him to identify a variety of factors. As a result, he obtained advice from people with experience on this type of engine, as well as the technical document, which we were admiring for its excellent condition.

Based on these experiences and documents, Oliver offered to examine how the bogies were secured. To fully understand this suggestion, I first had to remember that a bogie is a small trolley with 2 or 3 axles. They are used on a large number of coaches, wagons, railcars, and locomotives, because these rolling stocks are so long that they would struggle to take bends if the axles were directly attached to their chassis. Generally, bogies are fastened to the chassis using a pivot. On this particular railcar, they were secured in a different way. A metal cable is used to connect them to the chassis.

Oliver pulled on the cables in order to assess their condition. He immediately determined that these cables were not sufficiently taut. Tightening these cables would therefore be the second task undertaken by the two volunteers. A screw was provided for this purpose. As it had rarely been operated, and had been exposed to all types of weather, this screw proved particularly difficult to move. It was therefore necessary to clean the place where it was located. I removed the dirt. Then, Oliver used a wrench, pulled with all his strength and leaned on the wrench with all his weight, although unsuccessfully. Ian joined him to combine their efforts. However, even the two of them together failed to achieve any result.

Without stopping for a moment, Ian asked me to fetch a tube of the same type as that used to try and straighten the brake linkage in the locomotive shed. Once I had found the right tube, he joined me again. Using a vice, we squeezed the end of the tube. By changing the round section into an oval section, we widened the end of the tube. Back in the pit, under the railcar, this modified tube enabled us to extend the wrench handle. Once pushed into the modified tube, the wrench had a lever arm of almost two meters. However, with its extended handle, the wrench proved uneasy to manipulate. Each attempt to turn the screw resulted in hitting one of the pit sides. Ian and Oliver moved to a different tool arrangement. They finally managed to handle the extended wrench from the pit by sliding the long arm between the track and the chassis of the railcar (see Fig. 8.2). This time, the strength of the two volunteers was enough to turn the screw and tighten the bogie cables.

Fig. 8.2 Arranging the extended wrench (© Photo: C. Lejeune)

Approaching the pit, Ivo, the dispatcher, asked us if everything was going well. He weighed up the situation before giving us a piece of advice. He suggested that we should not overtighten the cables, so that the railcar could still handle the tightest bends on the network without any difficulty after the adjustment. Finally, Oliver, Ian, and Ivo managed to agree on the ideal tightening strength.

Follow-Up Maintenance

Similar conclusions can be raised from the bogie cables issue as from the brake linkage restoration. As in the first vignette, these non-professional "workers" proved able to invent, modify, and repair their own tools, using the same materials that were used to build the steam locomotives. The tube taken from a steam locomotive boiler was actually used as a lever arm to straighten the brake linkage and to turn the wrench, in order to tighten the bogie cables.

At first glance, the repair itself looks similar. The loose cable can be tightened with nothing more than an (extended) wrench and (significant) workforce. This restoration nevertheless presents slight differences. Unlike the bent rod, the slackening did not result from the derailment, but rather is the alleged cause of the derailment. The slackening itself was not caused by a specific incident but occurred gradually over time. Strictly speaking, tightening the cables is a part of the overhaul activities. While the former operation is more like typical repairs, the latter normally falls under maintenance activities.

In fact, the task of tightening the bogie cables was certainly undertaken regularly by the maintenance staff when this railcar was in regular service for the national company. While the locomotive was used for many years by an association of volunteers, these same cables were not tightened during regular maintenance work. The derailment was an alarm signal that drew the volunteers' attention to the fact that something was missing from the regular maintenance of the rolling stock. That's why a major overhaul was organized after the problem arose. In fact, the cables in question were not tightened enough and the volunteers' task was to conduct a kind of follow-up maintenance.

Rather than preventive, the overhaul here is retroactive and reparative. It is fairly similar to a repair, as the problem has already arisen. This suggests that actors make no great divide between repair and maintenance. Indeed, everyday activities of the shop seamlessly combine repair and maintenance operations. Both straightening the bent rod and tightening the loose cables can thus be considered as restoring the locomotive to a previous state.

As noted in the previous section, restoration implies fixing a problem caused by a breakdown. In many cases, the damage consists of something not only disfigured but also broken into many pieces. Restoration then requires reassembling what has been broken. Pieces are joined together with glue, solder, or filler metal, or can be tied together with a hose clamp or fastener. A substance, such as glue or solder, or a tie, like a hose clamp or nail, is supplied to hold pieces together, but no original component is replaced.

Straightening, gluing, and fastening are common activities involved in the restoration of a damaged device to its original state. Such fixes do not require in-depth investigations to identify the cause of the problem. Unless something is irreparably broken, actions can be taken as soon as the affected component is identified.

Breaks and Interruptions

As with the brake linkage, the intervention on the bogie cables rested upon Ian and Oliver's combined workforce. Of course, an extended lever arm was needed. But this strategy is not exclusive to locomotive maintenance. Car technicians also happen to use improvised tools such as a length of pipe to extend spanners (Dant 2005: 112, 128) or a tire lever to adjust a sliding van door (Dant 2010: §3.3). As noted for the previous vignettes, such a tool arrangement allows car technicians to conduct this kind of repair alone. However, while Ian was able to straighten the bent rod alone, here, Oliver's workforce was not sufficient to turn the extended wrench. A second member's workforce was required. Contrary to the first restoration, this repair justifies teamwork for material purposes. Indeed, locomotives have significantly heavier engines than cars and, in selected occasions, such as here, size and weight matter. But differences between the two vignettes go beyond material features.

In the first vignette, the restoration rested on the smooth coordination of the two volunteers. However, the second vignette's restoration benefited from additional resources. First, Oliver's social network provided him with access to useful technical documents about the tautness the cable should be adjusted to.[2] In parallel, Oliver asked another association to confirm the relevance of such a setting for use in tourism.

In addition, a third person joined the original scene, bringing unexpected advice that influenced the final adjustment. Such an intrusion is not unusual. Indeed, volunteers frequently interrupt what they are doing. In such occasions, they usually move to their closest co-worker. They stand around, attending to others' repairs. The observed person can ascribe a specific meaning to the attendance according to his or her status, seniority, authority, or self-reliance. A novice may dread a controlling gaze over his shoulder or expect help to be provided. In contrast, an experienced or confident worker expects his dexterity to be admired. Occasionally, when a repair proves to be difficult, some members might also demonstrate irritation over others watching them work.

Sometimes, the onlooker may commiserate or comment on what is done, just as Ivo did. By doing so, present volunteers indicate that they care about what others do, reinforce themselves as team members, and make themselves available for further help or advice. As for mere presence, comments may be seen as either constructive or critical. To avoid resistance, suggestions are formulated within a subjunctive mode, which stimulates cooperation (Sennett 2012: 23). How such sights and comments are interpreted depends not only on self-reliance and seniority, but also past relations. Although with equivalent standing in the hierarchy, Oliver and Ian took Ivo's piece of advice very differently.

Support Circles

A spectator like Ivo does not stay alone for long. Other volunteers join him. Together, they form a circle of onlookers around the activity. These circles resemble the way jazz players arrange themselves when one of them performs a solo improvisation. Their watchful listening manifests support and availability. Similar support circles form around Repair Cafés tables (Rosner 2014) or, in the industry, around workers demonstrating their mastery of brazing (Dodier 1993). Thanks to such gatherings, two to five members share knowledge; the spectators see what the others are doing and stay up-to-date by doing so. If a novice is present, the circle is an occasion to explain what kind of repair is being conducted, and how and why it is achieved.

Often, the attendance or the comments of viewers stimulates the observed person to interrupt what he or she is doing. An informal meeting then occurs, providing an opportunity to exchange tips, as when Ivo advises not to overtighten the bogie cable. These interruptions allow the volunteers to collectively adjust how repairs are conducted. When a discussion emerges, each of the participants explains what he or she is doing, how he or she plans to achieve it, and the difficulties he or she anticipates. Every participant shares advice on the others' activities.

Within a few hours of work, each volunteer has the opportunity to explain what he or she is doing half a dozen times and to collect at least as many alternate solutions. Some of these proposals are included in the next story, as alternative options to solve the problem. When one of the options is approved by many spectators, the volunteer tends to integrate it to his strategy and adapts his story accordingly (see Orr 1996 on "war stories" and their contribution to repair work).

In contexts ruled by scientific organization of labor, such as assembly lines or production factories described by Linhart (1981) and Roy (1952), these interruptions would be problematic. Instead, in a shop not governed by imposed pace or quota, a volunteer does not infringe on official rules when he interrupts what he or another member is doing.

Better, in a voluntary context, repeated breaks fulfill a productive role in the social organization of repair. Such a double interruption ritual provides mutual awareness. These social engagements allow for the adjustment of repair activities. What an outsider would consider as endless chatting or as unproductive loss of time represents, from the team perspective, an efficient way to coordinate collective activities.

The productive role of support circles should however not be overestimated. Not all breaks result in a support circle, nor do all support circles succeed in finely adjusting a collective repair. Some interruptions might represent a simple opportunity to rest, to spot trains, or to chat about the weather. Someone not involved in the task from the start might fail in taking stock of the situation. His or her suggestions are then likely to have already been mentioned. They may also be completely off topic, and thus prove to be useless. Contributions of this kind might exasperate the recipient. A volunteer may also react badly to a piece of advice, even given with the best of intentions. Although rare, some critical comments may also occur. As with any group, the collective of locomotive keepers is not paradise on Earth.

In brief, breaks and circles stimulate cooperation when they provide mutual awareness, availability, help, constructive comments, or learning. In contrast, irrelevance, critiques, contempt, pride, misunderstanding, admiration, and mere demonstration of dexterity hinder the ability to work together (Sennett 2012).

This second vignette exhibits additional layers of the collective shaping of repair. As with the restoration of the brake linkage, the bogie cables were not adjusted by an isolated individual. Instead, the restoration rested on the kinship of two close friends. People devoted to adjusting the bogie cable were endlessly consulting each other. Moreover, other members do not miss opportunities to intervene while the repair unfolds, such as Ivo did. On these occasions, support circles may orient the ongoing activity. The repair also benefited from assistance outside the organization. When required, the association can count on external networks, like employees from national railway companies or locomotive keepers from other associations. Therefore, neither the volunteers nor the association proved to be isolated (on extended networks of repair work, see Houston in this volume).

Replacing the Blower

Third vignette

> On a track close to the pit where Ian and Oliver were restoring the railcar, another member, Larry, was firing up a former coal mine locomotive. I temporarily left the pit and went to ask Larry how things were going.

When I approached the cab, I saw Larry facing a tall flame escaping from the firebox. He swore. "This is not supposed to happen," he explained. "Draught is insufficient."

In order to understand Larry's diagnosis, I had to remember my first contacts with the field, two years earlier, when I was explained how a steam engine functions (I will relate this episode in full detail in the next section). As an apprentice, I learned that a steam locomotive is built around a firebox, a tube boiler, and a smoke box.

To fire up a locomotive, the fireman starts by lighting the fire in the firebox, first using soaked cloths and then pieces of wood. Coal is provided afterwards. While burning, coal produces smoke and hot gases. These hot fumes travel along the tubes, which pass through a boiler filled with water. They gradually heat the water, which is then converted into steam and provides the power used to propel the locomotive. After the fumes produced by the firebox have heated the tubes, they come out in the compartment at the front of the locomotive, which is known as the smoke box. In the same way, after the pressurized steam has propelled the locomotive, it also ends up in the smoke box.

Smoke and steam are discharged from the locomotive through the chimney above the smoke box. As with open fires, the locomotive needs a draught if it is to function properly. The more easily the combustion gases and steam can escape from the smoke box, the better the draught. This evacuation ensures that the firebox produces sufficient heat, so that there is enough pressure inside the boiler to propel the locomotive.

The smoke box is fitted with a blower. The latter consists of a pipe pierced with very small holes, through which the pressurized steam escapes. It causes the gases contained in the smoke box to be blown towards the top of the smoke box and directed towards the chimney. This device therefore makes it possible to regulate the draught.

Throughout the day, Larry and his crew experienced problems related to draught. I personally witnessed these difficulties as they attempted to get the locomotive moving after it had suffered problems on a few journeys. A diesel shunter came to the rescue of this steam locomotive, which was not powerful enough to pull the train.

From mid-day and through lunch, the reasons for this failure were discussed with some of the volunteers, including Larry and Darin, who was to run the steam locomotive the next day. Teams that run steam locomotives take frequent breaks throughout their working day. As railwaymen are exposed to strong heat from the firebox, these breaks enable them to stay fully hydrated. But breaks devoted to drinking a bottle of water are also an opportunity to coordinate the crewmembers, who organize train journeys on the railway line, and the team that works in the workshop, located at the end of the line.

In this case, the diagnosis focused, from a very early stage, on the blower. This pipe pierced with small holes, through which steam escapes, is meant to direct the exhaust gases towards the chimney opening, at the top of the smoke box. Larry and Darin's diagnosis pointed to the material from which this blower is made. According to them, the metal used was not sufficiently heat-resistant, causing the blower to bend slightly as the temperature rose. If the blower bends, the steam no longer travels in the same direction and is transferred less towards the chimney. This may explain the reduced draught inside the smoke box, tubes, and firebox.

On that day, the locomotive struggled to fulfil its purpose. The next morning, I joined Darin who was cleaning the smoke box before the locomotive was fired up. The two of us were standing in front of the locomotive, which had its smoke box door open, and looked at the infamous blower, as well as the ends of the tubes that come in the smoke box.

Darin picked up a rasp and rubbed it on the blower, where he discovered an orange metal (Fig. 8.3). "That's it," he said. "It's copper." In his view, the conclusion was clear. The blower, made from copper, was unable to withstand high temperatures and directed the gases contained in the smoke box in a direction which was not precise enough to provide an adequate draught. I then asked Darin about the possible options. He considered the idea of using Inox cables to hold the blower in place. Over the next few weeks, the workshop made a new blower from steel (which is more heat-resistant) and replaced the original copper blower with this new device.

Replacement

Restoring or reassembling is not always possible. This repair differs from the two previous ones in that no incident or breakdown caused the problem. A diagnosis step was therefore needed. The blower did not suffer from deficient maintenance. It did not need to be restored. It had to be replaced. The need to replace may depend on the material. Pieces made

Fig. 8.3 Inspecting the blower (© Photo: C. Lejeune)

of rubber cannot be glued or soldered. Similarly, corrosion over time prevents restoration. Deformed bearings, corroded boiler tubes, fatigued fireboxes won't be restored or reassembled but rather have to be replaced.

In everyday life, when a fuse blows or a belt cedes, we buy spare parts. Hardware stores provide spare parts to replace used components. Replacement occupies a relevant space alongside restoration and reassembly. It does not necessarily require different skills than restoring or reassembling. However, tools and manpower do not suffice any more. Spare parts are needed. The availability of these spare parts depends on existing distribution networks.

Maintaining locomotives unavoidably involves replacing worn components. But a steam blower is more difficult to obtain than a light bulb. Given that spare parts are not available for decommissioned locomotives, railway enthusiasts build substitute components from scratch.

In his famous book *Zen and Art of Motorcycle Maintenance*, Robert Pirsig (1999: 457–458) reports how a welder reconstructs a used chain guard. With no spare part available, the welder uses steel filler rod as a material to construct the missing piece. Because his chain guard is made of a thin sheet of metal, Pirsig was skeptical about such a strategy and feared the welder would punch holes in it. For him, brazing would have been a more appropriate technique. He was thus anxious when he entrusted his chain guard to the old welder. However, he admired the welder's technique and complimented him on the "beautiful" result.

The welder's repair occupies an intermediate position between reassembling and replacement. The filler metal is normally used as glue to reassemble pieces. Here, the welder uses it as building material. He constructs the missing component with what is on hand at the moment.

Here, the replacement of the blower did not result from a breakdown, but consisted in a modification of material. Such a substitution contradicted my assumptions. When I started my fieldwork, I assumed that preserving the heritage of steam locomotives implies that the original materials and designs should remain intact. However, volunteers demonstrated no hesitation when an original blower had to be replaced with a new device, made of a material and with tools or techniques from our contemporary society. Substituting materials or applying modern techniques to legacy rolling stock is definitely not an issue for them. "We reconstitute antique engines, but not with antique techniques," Larry confided on 6 September 2014.

Innovating in Preserving

Unlike archaeologists, locomotive keepers do not attempt to recover historical labor methods. They are rather concerned with running vintage trains. Because replacing worn components is unavoidable and granted that substitute devices will never behave exactly like the original ones, reconditioning old machines requires flexibility. Facing multiple constraints, those concerned with railway heritage prioritize their criteria. Running functional steam locomotives prevails over strict preservation. The Inox cables and the new blower fall under such a compromise.

Comparing the blower to water tanks expands the understanding of their purposes. Nowadays, steam locomotive water tanks are not fastened with solid rivets, but welded. Modifying the appearance of the side tanks is nevertheless out of the question. Consequently, the shop team places ornamental rivets along the tanks to preserve the appearance of the steam locomotive. In this instance, the preservation of railway heritage combines the acceptance of new materials, techniques, and tools, while retaining the appearance of the old train. The renewal process of these steam engines entails making new things look old (Roux 2008: 73).

It is therefore likely that the blower modification is not an issue because it allows the locomotive to be functional without affecting the esthetics of the engine.

The replacement of the blower nevertheless did not completely solve the draught issue, which kept on preoccupying Ray, the association's head director. He was constantly thinking about it, even while busy in his position at an industry factory. He thought of a breakthrough on a weekday and immediately phoned the Work Training Center housed in the workshop. The instructor, Leo, is a member of the association. Ray asked Leo to check the smoke box ceiling of other locomotives, stationed in the museum and around the workshop. Leo reported that, contrary to the coal mine locomotive, most of them were fitted with a cone aimed at collecting exhaust gases and directing it towards the chimney. On the next Saturday, Ray proposed to forge such a device, called a petticoat pipe, and to insert it in the smoke box of the steam locomotive experiencing draught issues.

This episode exhibits that the collection of locomotives deteriorating around the shop fulfills a threefold purpose. First, some of the decommissioned locomotives will be reconditioned, although only a few of them will become operational. Second, other locomotives are preserved for their components; when an engine is dismantled, many pieces are likely to be reused (Harper 1987). Even perforated, a boiler tube can help, for instance, to straighten a bent rod or work as a wrench extension arm. But locomotives are not only collected for their components. They also constitute a collection of the wide variety of steam locomotive equipment. As such, they supply information on the original designs and can be used as a diagnostic tool when a problem occurs on a preserved machine. For instance, when the team tries to solve the draught problem of the small steam engine, inspecting smoke boxes of various engines reveals that a petticoat pipe was missing.

Heritage and Amnesia

When I returned a few weeks later and was told about this story, I asked Ian why the locomotive did not originally have this petticoat pipe. “It is quite possible that it used to have it,” he answered. He suggested that it might have been removed when the locomotive was disassembled and reassembled in the shop, perhaps during its last overall (two years before) or during another modification in the shop. The cone was possibly not put back again, due to it being considered unnecessary. Or it might have been so corroded that it was removed from the smoke box and thrown away with the firm intent to build a new one. Perhaps this plan was forgotten between two Saturdays. Or, less likely, the volunteer that took this decision might have left the organization in the meantime.

It is therefore uncertain whether the petticoat’s addition is a restoration or an innovation in itself. However, this uncertainty did not preoccupy the volunteers. They did not need to know whether the locomotive had this petticoat pipe when it was originally designed almost seventy years ago. It was enough for them to diagnose the problem and look at how it has been resolved in other locomotives, in order to adapt the device to the relevant engine. This work on a locomotive engine did not require them to know its precise history. The members care about efficiency rather than about the original design of the blower and the petticoat pipe. Perhaps such a doubt is easier to accept because, contrary to the water tank rivets, both devices are hidden in the smoke box, and thus will not affect the esthetics of the locomotive.

Tim Dant (2005) depicts a case similar to the replacement of the petticoat pipe. He describes how two technicians remove and replace a whole car engine. They arrange the removed parts in boxes of different sizes, on the ledges of the engine compartment and even “on the floor mixed with sockets and other tools” (ibid.: 119). Their concern in refitting the engine is so tense that they “inhabit” these bits. They “read” the way parts fit together and manage to reassemble the whole engine by trial and error, without any instruction manual.

Although similar, the petticoat issue made the reading a bit more complex. Because shop activities are shared among a lot of people and because a large amount of time had elapsed between the disassembly and the re-assembly, the flow of action was interrupted so that the location of the original petticoat pipe and even its existence became unknown. Ignoring whether the smoke box was originally provided with a petticoat pipe is the drawback of a volunteer team.

Finally, the draught issue made clear that preserving and running heritage locomotives requires prioritizing paradoxical purposes. While preservation suggests engines to remain as they were, operating rolling stock unavoidably causes wear and tear. Maintaining a heritage railway requires the combination of paradoxical features, such as being conservative and creative. This can be complex and puzzling, which may include trial and error, as the next vignette shows.

Cooling Hot Boxes

Fourth vignette

> The steam locomotive that encountered the blower issue reminds me of the first day I spent in the workshop, on 8 September 2012. When I arrived, I was assigned the task of cleaning the body of the same steam locomotive. Definitely, cleaning is a typical assignment for a newcomer (Crawford 2009: 75). Another volunteer, Luke, was firing up the locomotive while I was starting to clean the sides of the body. I was soon interrupted by other members who noticed that I was new and welcomed me to the shop. As they introduced themselves, some members gave pieces of advice. One of them provided me with a larger bucket and with a cloth in addition to the sponge I was using. Another counselled me to start with the top of the body, rather than the sides, given that the temperature of the boiler was gradually rising.
>
> From time to time, some visitors surrounded the locomotive. "Is it heating?" they shouted at me. A few of them started questioning me, giving advice, and concretely exhibiting the community of railfans. Wanting to make a good impression to my new colleagues, I remained devoted to what I was doing and did not jump off the locomotive to answer. This reaction showed that I was not yet used to support circles.

Given the heating process takes 3 to 4 hours to complete, Luke had many occasions to facilitate my initiation. We indeed took several breaks together. These breaks represented opportunities to start my apprenticeship. Luke was the first to explain how a steam engine functions (see the previous section). Concurrently with these factual explanations, the ongoing breaks were teaching me the underlying ritual of the double interruptions.

When the steam attained a sufficient pressure, Leo joined us in order to test the locomotive. These tests involved several manoeuvres in the railway yard (tracks and points) around the workshop. As each time a locomotive starts moving, the driver gives a whistle signal. Given it did not match a regular schedule of the line, such a sound sufficed as a call for people working in the shop. Some of them interrupted their activities to come and admire the steam locomotive in motion. Unfortunately, very quickly, Leo climbed down from the locomotive, pushed his arm under the bodywork and, with a tense expression, exclaimed "Hot box," meaning that the axle box was overheating.[3]

Discovering Lunch Talks

It was midday. The three of us interrupted our tests and joined the other volunteers for lunch. As the weather was mild, we brought out a few chairs, so that we could eat outside in the fresh air. While we ate, my two workmates talked about the results of the tests. The score of people present discussed possible solutions.

Lunch is of dramatic importance for the collective coordination of the team. At noon, all shop activities come to a halt while the team members gather to share a meal. In contrast with support circles, gathering only subsets of the team, lunch is an interruption shared by all the people present in the shop. Stories previously developed in the shop are expanded during this time. With everyone present, this is the common place to review current issues, list plausible causes of locomotive failure, contemplate possible solutions, and keep each other up to date. In doing so, the team collectively diagnoses problems. Lunch talks provide an opportunity to exchange tips and coordinate repairs.

That day, the lunchtime discussion was dedicated to a collective brainstorm on the hot box issue (Orr 1996). I then realized that everyone was aware of the problem, which frequently affects this small locomotive. In order to reduce overheating, the volunteers had already tested increasingly thicker types of oil. Working in the industry, Ray identified the thickest oil. Leo explained to me that this oil is so thick that it feels like velvet if you let it run between your fingers.

As using this thick lubricant did not appear to produce the desired effect, the team decided that more serious steps needed to be taken with the locomotive. A couple of weeks before I started my fieldwork, the locomotive was therefore returned to the workshop. It was positioned over a pit and lifted (i.e., its wheels were no longer touching the rails) so that it was possible to work directly on the axle boxes. While explaining, Leo showed me some pictures on his smartphone. Another member, Serge, shot these pictures in the shop and published it on his blog (see Figs. 8.4 and 8.5).

Ray commented on Fig. 8.5. Inside the boxes, the axle comes into contact with the rest of the chassis through a bearing. This bearing is a concave piece of bronze, whose bore is coated with an antimony

Fig. 8.4 Lifting the locomotive (© Photo: Serge Poncé)

Fig. 8.5 Engineer's blue (© Photo: Serge Poncé)

alloy. In order to detect places affected by excessive friction between the axle and the bearing, the rear axle was smeared with some Engineer's blue paste. As expected, there were many blue marks. The first solution envisaged smoothing the concave part, which comes into contact with the axle. This antimony part was therefore machined, in an attempt to reduce the friction that was causing the axle box to overheat. Readers interested in (smaller) bearings are invited to read Tom Dant's chapter in this volume.

Combined with the thick oil, smoothing the bearing was seen as a solution to the hot box problem. The tests conducted with me belonged to a series of tests over several weeks. That day's session was the first since the locomotive was brought back down onto the rails. Unfortunately, these tests proved inconclusive. The axle box was still overheating.

Insulating the Ash Pan

During discussion between the volunteers, another solution was proposed. Leo offered to insulate the ash pan under the firebox. His suggestion was based on the following reasoning: a locomotive's firebox is similar to an open fire. Red-hot coals are placed on a grate. The cinders created when they combust fall through the grate into an ash pan so that they can be easily removed. Due to the temperature inside the firebox, these cinders are very hot. They conduct heat to the sides of the ash pan. Because the locomotive is made of metal, this heat is likely to spread, via the chassis, to the axle boxes, which overheat as a result. Insulating the ash pan could therefore present a solution to the hot box problem.

A substantial amount of work would be required to complete Leo's solution before the end of the season. Members around the table were therefore divided about dedicating their efforts to an option whose outcome remained uncertain. However, Leo had another trick up his sleeve. As the instructor supervising the Work Training Center in the shop during the week, Leo offered to ask the trainees to insulate the ash pan. Indeed, on select occasions, their learn-by-doing training includes some tasks benefiting the locomotive restoration. That's what Leo and his team set about over the next few weeks.

This third step, after using thick oil and polishing the bearing, produced the desired effect: the axle boxes became less overheated. The final solution was actually based on a different diagnosis from the two previous steps. Using a thick oil and polishing the bearing were based on a diagnosis that focused essentially on the question of friction or rubbing. On the other hand, insulating the ash pan was based on a diagnosis that pointed to the thermal bridge created between the firebox and axles. This does not mean that the only cause of the axles overheating was that identified by the final diagnosis. In fact, it is likely that the two effects—friction and thermal bridge—had combined to result in the hot box.

Trial and Error Maintenance

These repairs were conducted in several stages. As Ray commented on 14 March 2015: "Much has to be tested before you find a solution. And, after you found it, you realize it is blindingly obvious." The first diagnosis led to the bearing being adjusted and a thicker oil to be used. As these measures failed to do the trick, the ash pan was insulated. It is interesting to note that actors were neither conducting conventional repairs, nor routine maintenance. In fact, the locomotive was modified when the ash pan was insulated. For several decades, it worked without this ash pan producing the heat that caused the axle box to overheat. In contrast with the railcar derailment, no accident or breakdown caused any changes that could have explained why it would become necessary to insulate the ash pan.

I only came to understand the reason for this step after asking members of the association. A few months before, the locomotive had been fully restored, which included an entirely new firebox.

Nowadays, replacing old steam locomotive parts does not rely upon the production and distribution networks originally developed by national railways. Deprived from these resources, keepers of decommissioned rolling stock collect and store components or learn how to create them on their own. However, maintaining a steam locomotive involves activities a bit more complex than replacing a blown fuse, building your own chicken coop, or even constructing your own chain guard by welding filler steel. A tube-boiler or a firebox requires industry-specific materials, tools, and techniques. Even highly skilled volunteers are not able to construct it on their own. For these reasons, the non-profit organization has established some industry partnerships. The firebox was renewed thanks to this kind of partnership. An industry factory constructed the new firebox following the current standards, using thicker and more resistant materials than those used seventy years ago.

The sides of the new firebox are thicker than the original model. With more heat-resistant sides, the locomotive firebox requires a hotter fire to provide the same temperature to the boiler, leading to unexpected consequences regarding the overheating of the rear axle boxes. In short, when the firebox became obsolete, the restoration of the locomotive became necessary, which subsequently led to the ash pan being insulated, as part of a domino effect. This also explains why a steel blower was required, while a copper one did the job perfectly for seventy years. This also gives additional clues as to why the volunteers seem unconcerned about the presence of an original petticoat pipe in the smoke box. They are likely to modify the original design because the renovation of the firebox was unavoidable, as are the consequences of this replacement. For the locomotive to be able to draw a train, there is no alternative other than adapting the smoke box organs.

Unavoidable Reconfigurations

The vignette illustrates that the insulation of the ash pan and the replacement of the blower were made necessary due to a previous renovation. It shows that a repair hardly leaves things unchanged.

Such a domino effect may be overwhelming. But people involved in preserving steam locomotives know this is an endless job. Throughout all seasons, rust will have to be removed, other components will have to be replaced and substitutes will cause new issues. However, far from discouraging the rail enthusiasts, these never-ending issues stimulate them; each season brings new challenges, representing opportunities to better understand how steam engines work.

The last two vignettes show that repairs are not restricted to restoration but can introduce novelty. The firebox replacement modifies—although not purposely—how the frame conducts heat. As a consequence, volunteers envisage adding Inox cables, fitting a new blower and insulating the ash pan. These modifications were probably all required for the same reasons: the temperature inside the locomotive is possibly higher than before, due to the new firebox. Interestingly, the work conducted on the blower and on the ash pan represent improvements to the original design or, at the very least, improvements to the condition of the locomotive when it was acquired by the people that currently use it. In both cases, repairs embody a reconfiguration.

Conclusion

The repair and the maintenance of vintage engines involve multiple strategies—restoring to a previous state, replacing a broken or worn device, and adapting the original design. These strategies not only contribute to a typology of repair activities. They moreover exhibit how repairs require the combination of apparently paradoxical features, such as preserving while using. As tinkerers, lay railwaymen keep used devices (for components and benchmarking) and are enterprising enough to create missing components on their own. This combination of conservatism and innovation fits Do-It-Yourself (DIY) independence. But unlike isolated tinkerers, lay railwaymen illustrate a deep social organization of repair and maintenance.

First, inside the shop, breaks, double interruption rituals, support circles, and lunch talks underpin the ongoing, dynamic, and mutual contacts among members and push them to endlessly reconsider and improve the ways repairs are conducted. This informal organization helps the team to be both productive and socially cohesive. Moreover, the shop can count on five external networks within other heritage railways, some people inside national railway companies, the two instructors at the Work Training Center, partners in the industrial sector, and the large numbers of railfans taking part in their activities. As noted by Matthew Crawford (2009: 27), this demonstrates support and contributes to the motivation of the people involved in preserving old engines.

This internal and external social networking helps to collectively maintain and repair heavy rolling stocks, as volunteers have been doing for half a century. They consequently foreshadow what Web 2.0 enthusiasts have been celebrating as an innovation of the twenty-first century—amateurs with meager professional background in the field joining their efforts in order to collectively perform complex tasks. As such, like hackers, lay railwaymen foreshadow Do-It-Together or Do-It-Ourselves (Frauenfelder 2011: 51) initiatives such as Repair Cafés, Fixit Clinics (Rosner 2014), FabLabs, and the Makers movement (Lallement 2015). I therefore hope my analysis might stimulate additional fieldwork on collective repair and comparisons of the underlying dynamics and peculiarities of such initiatives.

Finally, this ethnography on collective repair of steam locomotives is an invitation to reflect on our role in modern society. First, it testifies that people originating from varied milieus can genuinely cooperate. Second, locomotive restoration workers teach us something about the human ability to interact with artifacts (Dant 2005: 135). In a world dominated by programmed obsolescence, shall we throw away a defective device or are we going to repair or maintain it? Do we really care about objects? As such, this ethnography talks about our everyday life. On a broader perspective, studying repair also matters "just because modern society is now in urgent need of repair" (Sennett 2012: 199). Are we going to work at restoring the world as it was before the 2008 financial crisis (Lejeune 2011b) or are we going to reconfigure it?

Acknowledgements This chapter would not exist without Nicolas Dodier's inspiring sociology of machines. I am also indebted to Patrick Italiano who initially drew my attention to this field. I want to warmly thank all the volunteers at the *Three Valleys Steam Railway* that welcomed me, spent time to introduce me to their activities, and treated me as a peer. A special thank you to Serge Poncé for permission to reproduce the photographs he shot (Figs. 8.4 and 8.5). I am also grateful to Manon Bertha, Rachel Brahy, Tim Dant, Thomas Vangeebergen, an anonymous reviewer, and the editors of the volume for their comments on a previous version of this chapter. David Ward, Patrick Reichert and Charles Lee helped me with the English text. Of course, I take full responsibility for the mistakes or inaccuracies that remain within this chapter.

Notes

1. Frederick Gamst (1980: 59–60) reports how professional railwaymen deal with safety rules. Johan Sanne (2014) depicts how exceptions to safety rules are negotiated between the dispatcher and railway technicians maintaining the tracks.
2. As in car garages, manuals are rarely consulted, and only for standard figures
3. Frederick Gamst (1980: 88–90) provides an alternate ethnographic account of the hot box issue.

References

Crawford, Matthew B. 2009. *Shop Class as Soulcraft: An Inquiry into the Value of Work*. New York: Penguin.

Dant, Tim. 2005. *Materiality and Society*. Maidenhead: Open University.

———. 2010. The work of repair: Gesture, emotion, and sensual knowledge. *Sociological Research Online* 15 (3): 1–7.

Dodier, Nicolas. 1993. Les arènes des habiletés techniques. *Raisons Pratiques* 4: 115–139.

Frauenfelder, Mark. 2011. *Made by Hand: My Adventures in the World of Do-It-Yourself.* New York: Penguin.

Gamst, Frederick C. 1980. *The Hoghead: An Industrial Ethnology of the Locomotive Engineer*. New York: Holt, Rineheart and Winston.

Goffman, Erving. 1956. *The Presentation of Self in Everyday Life*. Edinburgh: University of Edinburgh Press.

Harper, Douglas. 1987. *Working Knowledge: Skill and Community in a Small Shop*. Berkeley and Los Angeles: University of California Press.

Lallement, Michel. 2015. *L'âge du faire. Hacking, travail, anarchie*. Paris: Seuil.

Lejeune, Christophe. 2011a. From virtual communities to project-driven mediated collectives. A comparison of Debian, Wikipedia and the Open Directory Project. In *Collaborative Search and Communities of Interest: Trends in Knowledge Sharing and Assessment*, ed. Pascal Francq, 10–20. Hershey: IGI Global.

———. 2011b. From normal business to financial crisis... and back again: An illustration of the benefits of Cassandre for qualitative analysis. *Forum Qualitative Sozialforschung/ Forum: Qualitative Social Research* 12 (1) (Art. 24): 19.

Lévi-Strauss, Claude. 1962. *La pensée sauvage*. Paris: Plon.

Linhart, Robert. 1981. *The Assembly Line*. Amherst: University of Massachusetts Press.

Orr, Julian. 1996. *Talking About Machines: An Ethnography of a Modern Job*. Ithaca and London: Cornell University Press.

Pirsig, Robert Maynard. 1999. *Zen and Art of Motorcycle Maintenance: An Inquiry into Values*. New York: HarperCollins.

Rosner, Daniela K. 2014. Making citizens, reassembling devices: On gender and the development of contemporary public sites of repair in Northern California. *Public Culture* 26 (1): 51–77.

Roux, Baptiste. 2008. *La poésie du rail. Petite apologie du voyage en train*. Paris: Transboréal.

Roy, Donald. 1952. Quota restriction and goldbricking in a machine shop. *American Journal of Sociology* 57 (5): 427–442.

Sanne, Johan M. 2014. Vulnerable practices: Organizing through bricolage in railroad maintenance. In *Vulnerability in Technological Cultures*, ed. Anique Hommels, Jessica Mesman, and Wiebe E. Bijker, 199–215. Cambridge and London: MIT Press.

Sennett, Richard. 2012. *Together: The Rituals, Pleasures and Politics of Cooperation*. New York: Penguin.

Thibault, Martin. 2013. *Ouvriers malgré tout. Enquête sur les ateliers de maintenance des trains de la Régie autonome des transports parisiens*. Paris: Raisons d'agir.

Part III

Politics

9

Negotiating Repair: The Infrastructural Contexts of Practice and Power

Christopher R. Henke

Introduction: Repair and the Negotiation of Infrastructural Orders

An office worker makes a call to the maintenance office at the university where she works as an administrative assistant. Her complaint? The office is too cold, and could someone check out the problem for her? When the mechanic arrives to investigate the office's ventilation system, he finds that there is "no problem"—the office temperature and the rate of airflow through the office vents are within the range of "normal" expected for that building and day. While the mechanic tried to use instruments and flashing numbers to convince the office worker that her workplace was not in need of repair, she still felt cold, and was frustrated by the lack of a straightforward fix for her complaint.

Another example: U.S. scientists working at Los Alamos National Laboratory built a nuclear weapons stockpile over decades, using

C. R. Henke (✉)
Colgate University, Hamilton, NY, USA
e-mail: chenke@colgate.edu

I. Strebel et al. (eds.), *Repair Work Ethnographies*,
https://doi.org/10.1007/978-981-13-2110-8_9

hundreds of tests to develop and refine their designs. In the 1990s, when the U.S. and many other nuclear powers signed the Comprehensive Test Ban Treaty, weapons scientists lost the ability to detonate nuclear devices as a part of their design process. This change not only impacted the work of these scientists—it also called into question the credibility of the "deterrent" effect given to the U.S. nuclear stockpile. Testing sent a very public message to other countries about the atomic capabilities of the nation, so how could weapons scientists, military brass, and politicians convey the threat of a U.S. nuclear arsenal without testing? In order to repair the damaged credibility of the nuclear threat, some supported an approach that would maintain and extend the life of the existing stockpile, while others argued for a new weapon design that would not rely on testing.

These examples, each taken from empirical case studies previously published by colleague Benjamin Sims and myself (Henke 2000; Sims and Henke 2008, 2012), emphasize the place of repair in diverse work settings, from an office to a global system of power and politics. I begin with these examples to emphasize the complexities of repair, where questions such as—is repair needed? how should something be fixed? who should do the work?—are negotiated among the people and the material infrastructures of diverse contexts. Scholars working in the field of infrastructure studies emphasize the embeddedness of these systems and how they are "'sunk' into…other structures, social arrangements and technologies," forming a crucial "part of human organization" (Star and Ruhleder 1996: 113; Star 1999: 380). Communication, transportation, energy, food, security: these structures are so thoroughly embedded in our daily life practices that we often take them for granted. And yet, though these structures are often invisible or transparent to us, a hidden basis of social order, they also often cause trouble. Breakdowns, glitches, and inefficiencies great and small are part of each of these systems, and just as we rely on infrastructures, we also rely on the work that it takes to maintain and repair them (Edgerton 2007; Graham and Thrift 2007; Graham 2010; Strebel 2011; Wallstein 2013; Jackson 2014; Denis and Pontille 2015). Though we typically think of repair as a moment of inconvenience or exasperation when a system fails to perform as expected, in this chapter I argue that repair is a fundamental feature of infrastructural orders, and that repair is all around us, all the time, in our everyday interactions with all sort of infrastructures.

Following work in organizational theory, I describe repair as a practice for negotiating order in contexts where heterogeneous elements come together to create complex social and technical systems (Strauss 1978; Fine 1984; Hughes 1983, 1989; Henke 2000). These "negotiated orders" involve interactions between people, tools, and infrastructures, where actors continually work out repair strategies to achieve their goals. Sometimes this repair is done by those we typically think of as repair workers—the mechanic, the plumber, or the tailor—but the view of repair I propose here is a more expansive conception, where repair is happening more often, and more people are doing it, than we might expect. In fact, repair is built into the fabric of everyday interactions, both with other humans and with the material infrastructures that support and shape our activities and interests; I focus on repair as a form of relational negotiation that mediates the connections between people, organizations and institutions, and materiality.

This vision of repair draws heavily on principles in ethnomethodology, and especially the way that repair has been defined through analysis of natural conversation between partners or small groups (Garfinkel 1967, 2002; Sacks et al. 1974; Schegloff et al. 1977; Schegloff 1986, 1992, 1997a, b; Jefferson 1987). Research in this field identifies specific practices of conversational repair, where the maintenance of interpersonal meaning is an ongoing achievement, continually monitored by actors who respond in the moment to situations of breakdown through the flow of language and understanding. I seek to extend that view of repair to the pervasive material structures of modernity. By focusing on repair practices, I emphasize a theory of infrastructures that is less... structured. Infrastructures often seem especially settled, fixed in place, and stable for long periods of time, but this apparent settlement is itself the product of a great deal of repair work. This perspective on infrastructural repair provides a contrast to the view of sociotechnical systems often emphasized in the field of science and technology studies (STS), where the metaphor of the "black box" highlights how knowledge claims and technologies become unquestioned matters of fact

(Latour 1987). Instead, the focus here is on how repair acts as a kind of invisible hand behind the stability of infrastructures.

This chapter explores the negotiation of material orders via repair in two key infrastructural contexts: the local and the institutional. The research vignettes above illustrate in brief the distinction between these levels of analysis, which are distinct in terms of scale more than in kind. At each scale I focus on negotiations over repair and questions about whether repair is needed and how to address it; these negotiations reveal important dynamics of practice and power, discourse and materiality. I support analysis of each scale with case studies developed through my own empirical research as well as other scholars' published accounts of infrastructural repair and politics. In the concluding section of the chapter I discuss the methodological considerations of studying repair via ethnography, given the centrality of ethnographic methods in studying repair in both my own work and in the literature more broadly.

Negotiating Local Infrastructures: The Relational Ontology of Repair

Where does repair come from? At first blush, repair is seemingly random: things decay, rust, and break down, and some means of fixing is required to make them whole and in good working order again. This is an entropic view of repair, where it is built into the structure of the universe ("And on the 8th day, God had to invent repair…"), and surely repair is often a reactionary process, responding to the inevitable influence of thermodynamics. But this approach assumes that repair, when needed, is relatively obvious and that the content or direction of repair activities are themselves taken for granted and made in response to straightforward breakdowns. The genesis of repair begs the question of an ontology, or how repair—and especially the perception of a *need* for repair—comes into being. In this section I argue that, though repair does come in part from physical processes of failure, repair is also negotiated through the interaction of humans, their material environment, and the discursive frames that embed infrastructural systems with meaning and interests. Breakdowns bring repair to our attention, but the ontology of repair is more complex than a response to material failure. Instead, repair work emerges from a negotiated order of materiality and human activity. Because infrastructures combine culture and materiality in ways that are hard to disentangle, the material basis of infrastructures make social relations seem hardened and static. But the theory of repair I argue for here emphasizes the dynamic aspects of established infrastructures. Thus, the "structure" in infrastructures is an ongoing accomplishment, negotiated through repair, as in the kinds of conversational repair studied by ethnomethodologists (Sacks et al. 1974; Schegloff et al. 1977).

My focus in this section is on the local negotiation of materiality and discourse in specific work settings, as described by a range of case studies on infrastructural repair. By local I refer to a scale where repair work on specific systems in particular places is accomplished through the negotiation of small groups of persons, sometimes just a solitary repair worker engaging with a local infrastructural order. Dant's analysis of materiality and the increasing complexity of this "material interaction" in modernity helps to conceptualize the role of infrastructures in local settings of practice (Dant 2005: 111; see also Edwards 2004). Our dependence on systems of technical control and mastery means that most aspects of our daily behavior are embedded in the stuff of materiality. Materiality provides a context for both enabling and limiting action; like culture provides a kind of "toolbox" of competencies that we draw upon to support our goals (Swidler 1986), material contexts provide both important affordances and constraints. In this way, material culture is more than the symbolic investments that we make in physical objects and settings. Using examples from his work on automobiles and their repair, Dant describes a threaded nut and bolt as "[artifacts that] embody the intentional actions of prior human beings," thereby building meanings and uses right into the form and structure of these objects (Dant 2005: 111–115). This point is affirmed by work in actor-network theory, which ascribes a kind of passive agency to materiality. Actors of all kinds, linked in networks of humans and things, create the basis for understanding materiality, including scientific facts and theories (Latour 1984, 1988, 1992; Callon 1986; Law 1987, 1994, 2002; Sayes 2014).

This networked complexity of infrastructures means that repair is implied in the very structure of sociotechnical systems, and that, when a basic breakdown occurs, structures provide hints at the need for repair and the form that repair will take. To return to Dant's example, if a bolt breaks off a machine, a repairperson most likely needs to get another bolt and replace it. At the same time, decisions about repair are often not this straightforward, and ascribing too much structural or network power to materiality may lead us to an overly deterministic theory of its influence. For example, in the most extreme version of actor-network theory, "culture" or "society" have no useful meaning, because everything is at base a network of actants, and these terms (as well as "knowledge" or "fact") are epiphenomena that we associate with extensive and robust actor-networks (Latour 1988, 1992, 1999). Determining the genesis of repair requires an approach that balances cultural and material aspects of our relationships with everyday systems, infrastructures, and workplaces. For this reason, I emphasize the *relational* ontology of repair, and especially the way that people, their material environment, and their interactions with each other, provide the local context for repair decisions and activities (Star and Ruhleder 1996). In prior work I have described this aspect as "people repair," to highlight the way that repair is about fixing peoples' orientation to specific settings (Henke 2000). Describing this process as relational repair includes fuller attention to material interaction as well as the way that actors in specific contexts negotiate meanings and actions around repair (Orr 1996). In the remainder of this section, I focus on two key aspects of relational repair: the role of the body (including emotional work) and local discourses of repair.

The Networked Body

Bodies have a paradoxical visibility and invisibility in the repair of infrastructure, and the invisibility of infrastructures themselves can be understood in part through the role of the human body. Watching repair in action, it is hard to miss the role of bodies and materiality—repair persons engage with materiality through their bodies and tools employed as extensions of bodily efforts and senses. And yet, the character of bodily engagement can often be mysterious to those who observe or depend on repair, as the very embodiment of knowledge and skill make them hard to "see."

Douglas Harper's *Working Knowledge* (1987) provides a vivid example of the importance of the body for repair. Harper focuses on the case of a mechanic named Willie, who owns a small repair shop in rural New York, fixing cars and farm equipment, often in exchange for favors, materials, and other resources necessary for surviving in an environment where winters are long and full-time employment is scarce. Willie is an important person in his community due to his expertise in *bricolage*, or the art of using materials at hand in creative and skilled ways. For Willie's customers, paying cash for a new car or tractor part is often difficult, and so Willie is a master of fabricating and improvizing solutions that are less costly and often more effective than replacing old parts with new ones. Harper emphasizes the importance of bodily knowledge for this work, as Willie must use craft knowledge of metalwork, welding, and reassembly to effect repairs. In one exchange, Harper focuses on Willie's expertise with automobile transmissions, and especially the skill required to disassemble and reassemble them; Harper notes that, as purely mechanical devices, transmissions should be relatively straightforward to repair, as they fit together in a manner governed by the logic of gears and illustrated through technical manuals. And yet: "Even in this most 'objective' of procedures it is the subtle play of force and pressure, the simultaneous movement of parts, and an evaluation of wear through the sensations of the fingertips that guide and control the process of work" (ibid.: 121). At one point, when discussing a particularly complex transmission, Willie describes the use of bodily sensations via hands and mind to understand the way it fits together, "just like your fingers got eyes" (ibid.: 126).

Willie's use of his fingers to feel and intuit his way through a transmission shows the deeply relational character of his work, and how he has built an intimate knowledge of transmissions into his body through engagement with material structures (Star and Ruhleder 1996: 113; Dant 2010). While "embodied" knowledge is a key aspect of his work, it is misleading to say that the body itself is the sole source of Willie's skill. Instead, I emphasize the "networked body" in repair work, where, "a body has become situated within a larger setting of activity, rather than simply internalizing a previously external skill" (Henke 2000: 63). A focus on the networked body highlights the history of material interaction between bodies and infrastructures, helping to explain the seeming mystery of bodily knowledge; knowledge is not just in the body but distributed through the stuff of particular machines and systems. This connection allows us to see the ontology of repair, because the relationship between bodily senses and material settings provides a key indicator when things need to be fixed—something looks wrong, smells bad, feels too hot or cold, or does not sound right. The sensation that something is wrong with local infrastructures is the genesis of repair, and this bodily ontology includes not just repair persons but also everyday users and inhabitants of infrastructures, because their bodies also often "know" when something is wrong, though they may not have the knowledge to more fully understand or repair the problem behind that sensation.

In a prior study focused on a group of physical plant mechanics working at a large university, I noted the role of the body in troubleshooting repair situations (Henke 2000). These mechanics frequently used their senses as a key tool for diagnosing problems on their campus and effecting repairs. A repair call typically began with a worker's bodily sensation that something was wrong with their work setting; for example, a common complaint involved an office that felt too hot or too cold, and a mechanic would be dispatched to investigate the call. Like Willie's description of working through a transmission using multiple forms of bodily and technical knowledge, the mechanics would employ their bodies, tools and diagnostic equipment, and an understanding of building infrastructures for energy, plumbing, and heating and cooling. In an interview with one mechanic, he described to me the use of an instrument for testing the rate of airflow through a duct, and how that tool allowed him to "look" inside the duct (ibid.: 64). Methodologically, I was able to observe this kind of work through observational and participant observational techniques, but some of the richest data about the networked body came through interviews where I asked the mechanics to tell me about their tools and situations in which they might use them.

In the example above, the mechanic opened up the back of his van, I set down a tape recorder, and he described several tools and instruments that he uses in his work related to the ventilation systems of the campus buildings. By focusing on tools, I was able to elicit several detailed examples where the mechanic used embodied senses in conjunction with instruments and building infrastructures to address repair problems, highlighting the relational connections between the body, materiality, and repair.

Discursive Repair—Talk at the Local Level

When breakdowns appear in a particular site, the ethnographer's body may not be among those that are first to notice; instead, those bodies more accustomed to the everyday sensations of a given place may alert the researcher through their talk. Again, Harper (1987) provides an outstanding example of how to conduct a study on repair that takes seriously the interplay of the body, materiality, and discourse; Harper employs a "photo elicitation" interview technique that uses images to elicit responses from an interview partner (see also Harper 2002). The embodied character of repair work may prove difficult for the researcher to follow in the moment. By taking photographs of Willie at work, Harper captures his repair work in action and can then use the images after the work is complete to explore Willie's explanations and meanings around the work.

Despite the possibility of capturing repair through images and narratives, talk about repair may complicate the ontological explanation I provided above, where repair appears when a body senses that something is out of order. What happens when bodies tell different stories, with competing suggestions for repair or views on whether repair is required at all? Because repair often opens up disruption and unease, breakdowns may reveal competing narratives of what is wrong, who is at fault, and what should be done. This is where relational repair becomes a truly negotiated order, with actors deploying potentially competing accounts of trouble and what to do about it. To return to the example presented above, where a repair worker is called to an office because someone finds the room too hot or too cold, a mechanic I interviewed claimed that changes in weather could shift the perceptions of office workers, where they might feel colder on a cloudy day. In this scenario, the mechanic questions the validity of a repair claim, and considers ways to reorient the relationship this office worker holds with his or her environment. Again, through the use of technology and instrumentation, the repair worker may be able to demonstrate an account of local reality to others in this setting. In this interview excerpt, a mechanic named Al described his use of an instrument that measures airflow through vents, such as one in the ceiling of an office:

> *Al*: [The flow meter] lets the customer know and see, that you're doing something and that there is actually [laughing] air coming out…and you can actually show 'em the reading. It seems to settle 'em down a lot of times. You know that's why in some of those labs you'll see, just a little strip of paper taped off of [the vent]—for years they've been having problems and [saying], "I'm not getting any air!" [mocking customer]. You'll hang that up there and take a reading and…let them see that there is air moving, and then you won't get the same repeat calls again.
>
> *CRH*: 'Cause they can see for themselves.
>
> *Al*: They can see yeah, and then they'll believe you. Where if you just go in and [hold] your hand [next to the vent] and go, "Yeah you're getting air out of here lady," they don't buy that, they wanna see something with some numbers flashing on it.

Al uses relational repair to readjust the office worker back to a sense of normalcy, using the flow meter as a means of producing consent around his own definition of the situation (Thomas 1923, 1928). In his view, repair of the building's ventilation system was not required in some situations, but instead the worker's own bodily sensations were mistaken and in need of repair (Henke 2000: 65–66). The gender dynamics present in this example, given Al's reference to a female customer, also point to the ways that gendered bodies might be seen as especially appropriate or inappropriate to sense the need for repair (see also Young and Rosner in this volume). The negotiation of repair is shot through with the potential for multiple categories of difference to shape the ontology of trouble in local work settings.

This connection between gender, discourse, and repair is also a key subject in Julian Orr's (1996) *Talking About Machines*, an ethnography focused on a group of Xerox technicians and their work repairing photocopiers. Orr details the "war stories" that the (largely male) technicians trade with each other, narratives that describe particularly troublesome and notorious repair situations (1996: Chapter 8). These stories serve multiple purposes in Orr's study, including the dissemination of knowledge about machines and repair, the socialization of new technicians, and the establishment of technicians' masculine identities vis-à-vis management. By telling war stories about repair, technicians can assert control over a job where unpredictability is common and relational repair is based in negotiation and improvization. While war stories may in some cases be a modest expression of a kind of oppositional identity to the institutions and managers that supervise repair workers and their activities, Orr also emphasizes the way that talk about repair is essential to getting repair work done, and that "the narratives are part of [workers'] practice" (ibid.: 143). In this way, relational repair helps corporations like Xerox to manage both the material functionality of its products but also the discursive frames that its workers and customers use to understand those products; relational repair creates organizational value and corporate profit.

A similar point can be made about the role of emotions in relational repair work. Hochschild's hugely influential concept, emotional labor, first introduced in *The Managed Heart* (1983), makes two key points about the role of emotion work, especially in interactive service work. First, emotions are essential to many aspects of work, and, second, management typically makes demands of workers to specify an emotional presentation of the self that ultimately serves the interests of the corporation, often more than the emotional needs and preferences of the worker (Hoschchild 1983; Wharton 2009). Workers at McDonald's are not necessarily smiling because they are genuinely happy to see their customers, but rather because they are required to put on a face of welcoming cheerfulness (Leidner 1993; Goffman 1959). Relational repair may involve emotional labor in terms of repair workers' own emotional faces, but also shifting the emotional states of the persons who are residents of or subject to infrastructures (Henke 2000; Dant 2010; see also Bovet and Strebel in this volume). As the mechanic Al, above, sought to use a flow meter to define an office's airflow as normal, he tried to push the office worker toward a different understanding and feeling of her work environment. At the local level, repair is continually negotiated through a dynamic exchange of talk, bodily skill and sensation, and material interactions.

Infrastructural Repair and the Negotiation of Power

Relational repair is a key moment for the negotiation of power. Conceptually, power is itself typically defined as relational, a system of control and resistance expressed through complex structures of interaction, and several of the examples that I discussed in the prior section pointed to the importance of power relations and hierarchies for understanding repair at local levels of analysis. For example, embodied knowledge, and the ability to employ that knowledge in specific material settings, can be seen as a form of power. Harper's (1987) *Working Knowledge* emphasizes Willie's social status in his community and Willie's ability to use his skills as a form of leverage that helps to enforce informal social contracts and relationships (ibid.: 184–198). Similarly, the example I described in the prior section, where a building mechanic used a hood flow meter to support a specific explanation regarding an office worker's physical environment, demonstrates his attempts to shape the nexus of infrastructure and discourse that give meaning to material settings (Henke 2000). Clearly, John Scott's (2001: 1) definition of social power as the "use of causal powers to affect the conduct of other participants in the social relations that connect them together" is partially achieved through the repair skills of Willie and Al (though I would add, "and material," to the relations referenced in Scott's definition; see also Foucault 1979, 1980a, b; Haraway 1991).

Overall, the relational character of repair means that systems of hierarchy and power present in infrastructures and related forms of human social organization are also always present via practices of repair. In this section I focus on this connection between infrastructures, power, and the scale of institutions and organizations, in order to emphasize the place of repair in negotiated orders. Note that the level of analysis treated in this section assumes a form of repair connected to but distinct from our everyday vision of a local worker like Harper's Willie, who fixes cars or tractors for his rural clientele. Repair of institutions or larger scale infrastructures still relies on embodied or technical knowledge, but that knowledge and skill may be deployed in the interest of others who control the direction and effects of repair more than the performance of the actual work itself. Just as for the case of repair at local scales of practice, this confluence of expertise and power raises an ontological question—where does the repair of power come from? Whereas localized repair emerges from the interaction of bodies, materiality, and discourse, the repair of power at institutional levels of analysis, while also based in these factors, focuses our attention on the intersection of expertise, commercial interests, and the state. When a group of engineers, scientists, politicians, or business owners identify a moment of repair, highlighting a crisis in a particular infrastructural system, we can ask where this comes from and how the call for repair connects with a system of material expertise and statecraft.

In this section I highlight two key factors that help to conceptualize the connections between material structures, repair, expertise and power. First, power relations are built into the structure of technologies and infrastructures; the influence of Foucault is critical for this point, as in his (1979) example of the panoptic power structure of the modern prison (see also Winner 1980; Zuboff 1988; Mukerji 1997, 2009; Scott 1998; Carroll 2006; Ureta 2015). The power embedded in infrastructures, in turn, is built on expertise developed through their construction and maintenance; modern systems of expertise, state power, and infrastructure develop in tandem and sustain each other through an ongoing confluence of interest. Second, and expanding on a point from the prior section on local talk about repair, discursive frames that define repair can be contested, meaning that repair itself becomes a site for divergent and even conflicting views of what form repair might take and whether repair is required at all. Because infrastructures are often interconnected and increasingly global in scope, the scale of these structures can complicate definitions of repair, seeming more or less radical at different points in the structure (Henke 2008, 2017; Sims and Henke 2008, 2012).

Creating and Repairing Infrastructures of Power

Chandra Mukerji's (1997, 2003, 2009) work on infrastructure and the formation of state power in early modern France provides a set of detailed historical examples of how structures built into the landscape, especially the gardens at Versailles and the Canal du Midi crossing southern France, capture local knowledge to consolidate formal expertise and create the infrastructural basis of power in the emerging nation state. Infrastructures as extensive as a 240 kilometer canal harness material power (in this case, water) on a massive scale, providing the means for a range of commercial and military interests that in turn support the credibility of state power (see also White 1995; Carse 2014). The fact that infrastructures such as canals are so fundamentally embedded in the landscape makes them seem permanent and fixed, but the material presence of these structures can obscure the systems of expert knowledge that are both created alongside infrastructures and help to sustain them through the cultivation of ongoing cadres of experts, their technologies, and their practices. In this way, infrastructures, and the institutions and organizations that are built up alongside them, form a kind of extended structure, reflecting an embedded system of power relations.

Once these extended infrastructures are put in place, expert knowledge can be deployed to negotiate the meaning of repair for them, acting as a resource for preserving and maintaining infrastructures over long periods of potential contestation and change. For example, in Henke (2007) I explored the ongoing threats to the city of New Orleans and its system of levees that protect the city from "normal" flood events but often fail to hold against periodic crises due to heavy rains and hurricanes, such as the 2005 Katrina disaster. Death and damage due to a strong storm like Hurricane Katrina was widely predicted many years before the events of 2005, and yet the inadequate levee system was not upgraded or replaced with a superior system. The key to understanding this failure is the embedded interests that created and maintained the levee system, a structure that largely reflected the stakes that elite landowners had in maximizing commercially lucrative land in the city versus a more land-intensive system of water control. Once the levee system was established, its material structure acquired a kind of inevitability, but one tied to a specific set of investments. The levee structure, in turn, reflected the expertise and interests of the U.S. Army Corps of Engineers, who continue to be a critical factor in supporting not only New Orleans' flood control infrastructures but also widespread infrastructures throughout the United States, constituting an enormous reserve of expert knowledge and skill for maintaining structures of power.

I examined a similar system of state-sponsored expertise in Henke (2008), a study of university and industry relationships in California's farm industry. California supplies much of the U.S. with fresh fruit, produce, nuts, and other crops that grow well in the State's Mediterranean climate. The intensity of production in California's farm industry, however, is contingent on a set of infrastructures that supply its growers with a diverse set of resources including water, labor, and expertise. A key source of this expert knowledge and practice is provided through the state via the Cooperative Extension service, a system of agricultural expertise supported by the U.S. Department of Agriculture, the University of California, and specific counties where Cooperative Extension "farm advisors" are based. Similar systems of agricultural science, geared toward the conditions and problems faced by growers in specific locales, were established in every U.S. State in the early twentieth century (Danbom 1979; Henke 2008: Chapter 2).

While much of the work of these advisors focuses on advice for practical matters such as how to treat an insect pest of a particular crop, farm advisors have also provided key support for the farm industry in times of more systemic crisis. For example, during World War Two, industry and state interests identified a farm labor "problem" due to the deployment of military forces for the war effort and the rise of industrial manufacturing in urban areas. This problem was actually a continuation of a decades-long struggle between farm industry interests and labor unionists seeking to organize the largely migrant sources of farm labor that growers depended on to grow and harvest their crops (Daniel 1982). Encouraged by industry pressure on the United States Department of Agriculture, Cooperative Extension acted as a labor contractor for the farm industry during the war years, marshaling workers—in many cases recruited from Mexico—to support California's farm fields. These efforts, in turn, required the creation of new systems of organization and practice to train workers and prepare them for labor with specific crops. For example, Fig. 9.1 depicts an, "Emergency Farm Labor Leaflet," created and distributed by the Cooperative Extension service, which describes the specific bodily movements and techniques required for harvesting sugar beets. Beet sugar was an important foodstuff during the war years, but sugar was also used in the manufacture of plastics and explosives, and so production of sugar was a key military imperative. Thus, the body, by harvesting sugar beets, was connected to a systemic form of repair that supported both farm industry and state interests in a time of global warfare. While this moment of "crisis" might have been a time for a more collaborative approach to labor relations among industry and union interests, instead the state, via institutionalized forms of expertise, provided large growers with the means to continue control over this food infrastructure and consolidate their power with respect to labor. Indeed, the repair mechanisms developed by Cooperative Extension during the war years continued for decades after the conclusion of the conflict (Henke 2008: Chapter 4).

The kind of direct observation and participant observation that I described in the prior section, when discussing Harper's work with Willie or my own work with the building mechanics, may also work for examining the scale of institutions such as those treated in this section. At the same time, the power structures embedded in infrastructures may be difficult to observe in the same way; for this reason, cultural, and historical analysis may help researchers studying repair to see how repair shapes infrastructures over long periods of time. In this way, complex institutional systems like a farm industry or a nuclear weapons complex can be seen as subject to pressures for both maintenance and transformation over longer periods of analysis, and at multiple scales.

HOW TO TOP AND LOAD SUGAR BEETS.

Very successful systems have been worked out for harvesting sugar beets by hand. Crews handling the largest tonnage with the least effort have developed a method which does away with all useless motions. The following outline shows one of the best methods.

The beets should be plowed in units of 16 rows from one direction. The toppers should work facing in a direction opposite to the way the plow traveled. Due to the slight tilt the plow has given, the beets will lift out easily.

Eight men work as a topping crew. Use the standard beet knife with the hook. Keep the knife sharp and the hook pointed.

No. 1. Stick the point of the topping knife into the beet on the side away from you, deep enough to lift the beet.
No. 2. On a medium sized beet, cut the top off, removing all green and brown stems. Do not lower tonnage by making a low or crooked cut.
No. 3. Very large beets with a pointed top should be topped off squarely.

Fig. 9.1 How to Top and Load Sugar Beets (University of California 1944)

Discursive Repair: Maintenance Versus Transformation

If power can be embedded in infrastructures, providing a kind of stability via the repair efforts of state experts, industry interests, and regimes of knowledge, as illustrated in the prior examples, this property of infrastructures begs the question of just how "negotiated" these structures truly are. Indeed, my descriptions of these examples underemphasize the negotiations among scientists, engineers, industry players, and representatives of the state, making them appear as a relatively easily interlocking set of allied interests, when in fact these groups often frame paths for repair in divergent ways. How do we understand negotiated orders in a context where the long-term stability of infrastructures is also apparent in their history and material and institutional structures? Clearly infrastructures evolve and change over time, and repair is a key part of this structural drift, but the interests embedded in infrastructures also often push toward a particular form of order. To help further theorize this balance between order and change, I have developed, along with colleague Benjamin Sims, two broad categories for conceptualizing paths for repair: repair as maintenance and repair as transformation (Henke 2007, 2008, 2017; Sims and Henke 2008, 2012).

Repair as maintenance seeks to stabilize existing social and material relations with respect to a given infrastructure; repair as transformation focuses on larger changes meant to more fully reorder infrastructure. These two categories of repair point to the importance of power relations as expressed through the social and material order of infrastructures. Typically, those invested in and benefiting through the existing structure of a system will prefer a repair as maintenance approach. Repair as transformation is more often proposed by those critical of established infrastructures, or in times of repair crises, where the very existence of a structure is called into question. While it may seem obvious that those with material interests in a given infrastructure will hold a preference for repair as maintenance versus a more radical transformation, the term "maintenance" should not be automatically be associated with a more modest set of repair activities and resources; in fact, what I describe here as a maintenance-based approach may actually be more costly and resource-intensive than a more transformative change.

For example, in the case of the New Orleans levee system, described above, the ongoing investment in the levee system represents a huge expenditure of capital, organization, and expertise, despite the continued risk of flooding. And yet that system does not fundamentally alter the network of interests that benefit from the investment in levees for flood control (Henke 2007).

This distinction between maintenance and transformation is especially helpful for showing the discursive politics of repair and divergent arguments that actors may make regarding the repair of infrastructures. For example, and to return to the example I briefly described in the introductory section, Sims and Henke (2008, 2012) explore the challenges faced by the U.S. nuclear weapons complex after the early-1990s test ban treaty that prevented weapons scientists and designers from testing nuclear weapons. In the absence of nuclear testing, scientists and administrators sought new methods of both developing weapons as well as securing the credibility of the existing nuclear stockpile in the U.S. arsenal. Ultimately, two key strategies emerged to cope with the loss of testing, including the Stockpile Stewardship program, which sought to preserve existing knowledge about weapons design primarily through computer simulations meant to virtually test the stockpile. Another program, the Reliable Replacement Warhead (RRW), instead focused on the creation of a new warhead design that would not require testing to develop and which would in turn have a very long shelf life as part of the U.S. arsenal.

In each case, the proposed solutions were meant to repair the credibility of both the weapons as well as the system of scientific expertise behind them. However, the Stockpile Stewardship Program, by arguing for a departure from the design of new weapons and major investment in new computing infrastructures, alienated the core weapons design community and was seen as a relatively transformative repair in terms of the discourses and institutions of the longstanding weapons development process. Though the RRW program seemed on face value to be more transformative, since it championed the development of a new weapons system, discursively and institutionally the program focused on the kinds of knowledge and skills more central to the older bomb design and testing process. This approach appealed to the core weapons design community since it seemed to preserve their autonomy and authority through a new weapons development regime (Sims and Henke 2012). These two case studies of technical, cultural, and institutional change in the face of a radically changed set of political and practical set of circumstances for work in the weapons complex demonstrate how, "repair may serve to reveal hidden differences," among the key stakeholders tied to a complex set of infrastructures (Sims and Henke 2012: 340).

Methodological Conclusion: Seeing and Representing Repair Through Ethnographic Methods

The examples of repair that I have discussed in this chapter come from researchers working largely through ethnographic methods, broadly conceived. If ethnography is defined as the close study of culture, especially situated cultural practice, then it makes sense that ethnographic methods are especially appropriate for studying infrastructural repair. By defining repair as a practice for maintaining and negotiating order in infrastructural settings, I have emphasized the importance of materiality, practice, discourse, and power for understanding how repair is used to maintain infrastructures. Especially at a local level of study, ethnography provides a useful set of tools for "seeing" repair in action. In this concluding section, I wish to elaborate the role of ethnography for studying repair and address some of the challenges of studying materiality via ethnographic methods. Ultimately, a broad approach to ethnography, and especially one that incorporates historical methods for a multi-scalar and multi-method approach to studying repair, can provide a rich view of this set of cultural and material practices.

Earlier in this chapter I discussed the challenges of theorizing materiality, and especially the balance between the hardness or agency of material infrastructures and the flexible interpretations that human actors may give to them. Traditionally, the social sciences have had mixed results studying materiality. Sociology, for example, has long been reticent to incorporate the physical characteristics of the human body, especially in analyses of inequality, for fear of perpetuating gender- or racially-based stereotypes about genetic and hereditary difference. Similarly, the rise of environmental sociology in the 1970s and 1980s was based in part on a critique of the "human exceptionalism" inherent in mainstream sociological analyses, where humans were treated as though they did not exist within a system of material resources and constraints (Catton and Dunlap 1978; Dunlap and Catton 1994). This exceptionalism tends to influence many handbooks on ethnographic methods, and most of these guides do not treat materiality or material culture explicitly. Instead, the context for research is described as a more immaterial "setting" or "scene" where the ethnographer works to gain rapport and learn local customs and languages (Emerson et al. 1995: Chapter 4; Schensul et al. 1999: 96–97; Bernard 2011: Chapter 12). Typically this setting is not conceptualized in terms of its materiality, though the inclusion of maps and map-making are sometimes recommended as a way of laying out the physical space in which activity occurs (Schensul et al. 1999: 105–111; Murchison 2010: Chapter 9). Material infrastructures and systems are not usually included in lists of potential topics and concepts for which ethnographers should be especially attentive (though see Tilley (2001) for a review of ethnography and material culture). In fact, when describing the role of "structure," most handbooks on ethnographic methods mean organizational and discursive structures, and not the influence of materiality.

The materiality of infrastructures and technical systems are embedded in networks of culture, discourse, and power, and ethnography is an especially appropriate set of methods for exploring these aspects of repair. Ethnography's attention to the role of the mundane and taken-for-granted beliefs and rituals of culture provides an important set of assumptions about what one should be looking for when studying the interaction of humans and their material culture. Note that I take a broad view of ethnography here, one that includes the traditional methods associated with ethnographic work, such as participant observation and interviews, but also historical-comparative, and discourse analysis methods, as each of these is very appropriate for a deep reading of culture. This diversity of techniques also makes ethnographic work especially appropriate for research at multiple levels of analysis, including the focus on local practice and institutionalization that I have analyzed in this chapter. Though the former is the level of analysis where repair has most typically been studied—and where repair as we imagine it in everyday life is located (fixing a car or patching a wall)—many systems and institutions important for understanding repair extend beyond the local and connect the globe in a complex network of structures with ongoing calls for repair (see Houston in this volume). Seeing repair as the long-term maintenance of power structures, especially through the cultivation of expertise, may require the ethnographer to shift his or her focus from local work settings to historical or organizational levels of analysis.

These two levels of analysis point to key advantages of ethnographic methods for studying repair. First, because repair is so connected with a set of skillful practices that are often embedded in experience, participant observation is an excellent way for the ethnographer to absorb, at least in part, the movements, procedures, and techniques that repair workers use to maintain and restore material objects and settings. Second, attention to the discursive practices that repair persons use to negotiate and frame the process of repair, often in interaction with other persons in the same setting, emphasizes the role of talk-based interaction in repair (Orr 1996). Combining these first two points, ethnographic methods allow the analyst to see the materiality of discourses themselves, as actors negotiate the meanings and literal structures that frame objects of repair. Third, because ethnographic methods typically involve a long-term commitment to a group or site, this longitudinal approach allows researchers to see the contingency of structures through repair practices; rather than seeing repair as a temporal disjuncture in social and material life, a long-term period of observation can help show how systems that seem "black boxed" are maintained through continual repair. In particular, an approach to ethnography that includes historical analysis provides a method for detailing the long-term stability and change of infrastructures, including the groups, interests, and resources that shape these processes (Mukerji 1997, 2009; Henke 2008).

On a more cautionary note, ethnographers must also be reflective about the politics of representation inherent in their methods, and how these politics impact the study of repair. Over the last several decades, ethnography has been thoroughly critiqued as a method of studying culture. Questions about identity, power, and representation challenge the epistemic value of ethnography and resituate the ethnographer from a disinterested and neutral observer of social settings to a subjective and self-interested meddler (or worse) (Clifford and Marcus 1986; Clifford 1988; Said 1989; Stocking 1991; Wolf 1992; Winddance Twine and Warren 2000). Though the harshest critiques call for the end of ethnography itself, the method has survived and even thrived in its traditional disciplines of anthropology and sociology as well as newer interdisciplinary fields, including especially science and technology studies. These critiques of ethnography as a research method mean that ethnographers studying infrastructural repair must attend to their own roles in the interpersonal and institutional power dynamics that they enter into when recording and writing about repair.

References

Bernard, H. Russell. 2011. *Research Methods in Anthropology: Qualitative and Quantitative Approaches*, 5th ed. Lanham, MD: Altamira Press.

Callon, Michel. 1986. Some elements of a sociology of translation: Domestication of the scallops and the fishermen of St. Brieuc Bay. In *Power, Action, and Belief: A New Sociology of Knowledge?* ed. John Law, 196–233. London: Routledge, Kegan, and Paul.

Carroll, Patrick. 2006. *Science, Culture, and Modern State Formation*. Berkeley: University of California Press.

Carse, Ashley. 2014. *Beyond the Big Ditch: Politics, Ecology, and Infrastructure at the Panama Canal*. Cambridge, MA: MIT Press.

Catton Jr., William R., and Riley E. Dunlap. 1978. Environmental sociology: A new paradigm. *The American Sociologist* 13 (1): 41–49.

Clifford, James. 1988. *The Predicament of Culture: Twentieth-Century Ethnography, Literature, and Art*. Cambridge, MA: Harvard University Press.

Clifford, James, and George E. Marcus. 1986. *Writing Culture: The Poetics and Politics of Ethnography*. Berkeley: University of California Press.

Danbom, David B. 1979. *The Resisted Revolution: Urban America and the Industrialization of Agriculture, 1900–1930*. Ames: Iowa State University Press.

Daniel, Cletus E. 1982. *Bitter Harvest: A History of California Farmworkers, 1870–1941*. Berkeley: University of California Press.

Dant, Tim. 2005. *Materiality and Society*. New York, NY: Open University Press.
———. 2010. The work of repair: Gesture, emotion, and sensual knowledge. *Sociological Research Online* 15 (3). http://www.socresonline.org.uk/15/3/7.html.
Denis, Jérôme, and David Pontille. 2015. Material ordering and the care of things. *Science, Technology and Human Values* 40 (3): 338–367.
Dunlap, Riley E., and William R. Catton. 1994. Struggling with human exemptionalism: The rise, decline, and revitalization of environmental sociology. *The American Sociologist* 25 (1): 5–30.
Edgerton, David. 2007. *The Shock of the Old: Technology and Global History since 1900*. New York, NY: Oxford University Press.
Edwards, Paul N. 2004. Infrastructure and modernity: Force, time, and social organization in the history of sociotechnical systems. In *Modernity and Technology*, ed. Thomas J. Misa, Philip Brey, and Andrew Feenberg, 185–225. Cambridge: MIT Press.
Emerson, Robert M., Rachel I. Fretz, and Linda L. Shaw. 1995. *Writing Ethnographic Fieldnotes*. Chicago, IL: University of Chicago Press.
Fine, Gary Alan. 1984. Negotiated orders and organizational cultures. *Annual Review of Sociology* 10: 239–262.
Foucault, Michel. 1979. *Discipline and Punish: The Birth of the Prison*. New York: Random House.
———. 1980a. *Power/Knowledge: Selected Interviews and Other Writings, 1972–1977*. New York: Pantheon Books.
———. 1980b. *The History of Sexuality, Volume 1: An Introduction*. New York: Vintage.
Garfinkel, Harold. 1967. *Studies in Ethnomethodology*. Englewood Cliffs, NJ: Prentice Hall.
———. 2002. *Ethnomethodology's Program: Working Out Durkheim's Aphorism*. Lanham, MD: Rowman and Littlefield.
Goffman, Erving. 1959. *The Presentation of the Self in Everyday Life*. New York: Doubleday.
Graham, Stephen (ed.). 2010. *Disrupted Cities: When Infrastructure Fails*. New York, NY: Routledge.
Graham, Stephen, and Nigel Thrift. 2007. Out of order understanding repair and maintenance. *Theory, Culture & Society* 24 (3): 1–25.
Haraway, Donna J. 1991. *Simians, Cyborgs, and Women: The Reinvention of Nature*. New York: Routledge.

Harper, Douglas. 1987. *Working Knowledge: Skill and Community in a Small Shop*. Berkeley: University of California Press.

———. 2002. Talking about pictures: A case for photo elicitation. *Visual Studies* 17 (1): 13–26.

Henke, Christopher R. 2000. The mechanics of workplace order: Toward a sociology of repair. *Berkeley Journal of Sociology* 44: 55–81.

———. 2007. Situation normal? Repairing a risky ecology. *Social Studies of Science* 37: 135–142.

———. 2008. *Cultivating Science, Harvesting Power: Science and Industry in California Agriculture*. Cambridge: MIT Press.

———. 2017. The sustainable University: Repair as maintenance and transformation. *Continent* 6: 40–45.

Hochschild, Arlie Russell. 1983. *The Managed Heart: The Commercialization of Human Feeling*. Berkeley: University of California Press.

Hughes, Thomas P. 1983. *Networks of Power: Electrification in Western Society, 1880–1930*. Baltimore, MD: Johns Hopkins University Press.

———. 1989. The evolution of large technological systems. In *The Social Construction of Technological Systems: New Directions in the Sociology and History of Technology*, ed. Wiebe. E. Bijker, Thomas P. Hughes, and Trevor J. Pinch, 51–82. Cambridge: MIT Press.

Jackson, Steven J. 2014. Rethinking Repair. In *Media Technologies: Essays on Communication, Materiality, and Society*, ed. Tarleton Gillespie, Pablo J. Boczkowski, and Kirsten A. Foot, 221–239. Cambridge: MIT Press.

Jefferson, Gail. 1987. On exposed and embedded correction in conversation. In *Talk and Social Organisation, Intercommunication*, ed. Graham Button and J. R. E. Lee, 86–100. Philadelphia, PA: Multilingual Matters Ltd.

Latour, Bruno. 1984. *The Pasteurization of France*. Cambridge, MA: Harvard University Press.

———. 1987. *Science in Action: How to Follow Scientists and Engineers Through Society*. Cambridge, MA: Harvard University Press.

———. 1992. Where are the missing masses? The sociology of a few mundane artifacts. In *Shaping Technology/Building Society*, ed. Wiebe Bijker and John Law, 225–258. Cambridge: MIT Press.

———. 1999. *Pandora's Hope: Essays on the Reality of Science Studies*. Cambridge, MA: Harvard University Press.

Latour, Bruno [Jim Johnson]. 1988. Mixing humans and nonhumans together: The sociology of a door-closer. *Social Problems* 35: 298–310.

Law, John. 1987. Technology and heterogenous engineering: The case of portuguese expansion. In *The Social Construction of Technological Systems: New Directions in the Sociology and History of Technology*, ed. Wiebe. E. Bijker, Thomas P. Hughes, and Trevor J. Pinch, 111–134. Cambridge: MIT Press.

———. 1994. *Organizing Modernity*. Cambridge, MA: Blackwell.

———. 2002. *Aircraft Stories: Decentering the Object in Technoscience*. Durham, NC: Duke University Press.

Leidner, Robin. 1993. *Fast Food, Fast Talk: Service Work and the Routinization of Everyday Life*. Berkeley: University of California Press.

Mukerji, Chandra. 1997. *Territorial Ambitions and the Gardens of Versailles*. New York: Cambridge University Press.

———. 2003. Intelligent uses of engineering and the legitimacy of state power. *Technology and Culture* 44 (4): 655–676.

———. 2009. *Impossible Engineering: Technology and Territoriality on the Canal Du Midi*. Princeton, NJ: Princeton University Press.

Murchison, Julian M. 2010. *Ethnography Essentials: Designing, Conducting, and Presenting Your Research*. San Francisco, CA: Jossey-Bass.

Orr, Julian E. 1996. *Talking About Machines: An Ethnography of a Modern Job*. Ithaca, NY: Cornell University Press.

Sacks, Harvey, Emanuel A. Schegloff, and Gail Jefferson. 1974. A simplest systematics for the organization of turn-taking for conversation. *Language* 50: 696–735.

Said, Edward W. 1989. Representing the colonized: Anthropology's interlocutor's. *Critical Inquiry* 15: 205–225.

Sayes, Edwin. 2014. Actor—Network Theory and methodology: Just what does it mean to say that nonhumans have agency? *Social Studies of Science* 44 (1): 134–149.

Schegloff, Emanuel A. 1986. The routine as achievement. *Human Studies* 9: 111–151.

———. 1992. Repair after next turn: The last structurally provided defense of intersubjectivity in conversation. *American Journal of Sociology* 97: 1295–1345.

———. 1997a. Practices and actions: Boundary cases of other-initiated repair. *Discourse Processes* 23 (3): 499–545.

———. 1997b. Third turn repair. In *Towards a Social Science of Language: Papers in Honor of William Labov, Volume 2: Social Interaction and Discourse Structures*, ed. Gregory. R. Guy, Crawford. Feagin, Deborah Schiffrin, and John Baugh, 31–40. Amsterdam: John Benjamins.

Schegloff, Emanuel A., Gail Jefferson, and Harvey Sacks. 1977. The preference for self-correction in the organization of repair for conversation. *Language* 53: 361–382.

Schensul, Stephen L., Jean J. Schensul, and Margaret D. LeCompte. 1999. *Essential Ethnographic Methods: Observations, Interviews, and Questionnaires.* Walnut Creek, CA: Altamira Press.

Scott, James C. 1998. *Seeing Like a State: How Certain Schemes to Improve the Human Condition Have Failed.* New Haven, CT: Yale University Press.

Scott, John. 2001. *Power.* Malden, MA: Blackwell.

Sims, Benjamin, and Christopher R. Henke. 2008. Maintenance and transformation in the U.S. Nuclear Weapons Complex. *IEEE Technology and Society* 27: 32–38.

Sims, Benjamin, and Christopher R. Henke. 2012. Repairing credibility: Repositioning nuclear weapons knowledge after the Cold War. *Social Studies of Science* 42: 324–347.

Star, Susan Leigh. 1999. The ethnography of infrastructure. *American Behavioral Scientist* 43 (3): 377–391.

Star, Susan Leigh, and Karen Ruhleder. 1996. Steps toward an ecology of infrastructure: Design and access for large information. *Spaces: Information Systems Research* 7 (1): 111–134.

Stocking, George W. (ed.). 1991. *Colonial Situations: Essays on the Contextualization of Ethnographic Knowledge.* Madison: University of Wisconisn Press.

Strauss, Anselm L. 1978. *Negotiations: Varieties, Contexts, Processes, and Social Order.* San Francisco, CA: Jossey-Bass.

Strebel, Ignaz. 2011. The living building: Towards a geography of maintenance work. *Social and Cultural Geography* 12 (3): 243–262.

Swidler, Ann. 1986. Culture in action: Symbols and strategies. *American Sociological Review* 51: 273–286.

Thomas, William I. 1923. *The Unadjusted Girl.* Boston, MA: Little, Brown.

———. 1928. *The Child in America.* New York: Knopf.

Tilley, Christopher. 2001. Ethnography and material culture. In *Handbook of Ethnography*, ed. Paul Atkinson, Amanda Coffey, Sara Delamont, John Lofland, and Lyn Lofland, 258–272. Thousand Oaks, CA: Sage.

University of California, Cooperative Extension. 1944. "How to Top and Load Sugar Beets".

Ureta, Sebastian. 2015. *Assembling Policy: Transantiago, Human Devices, and the Dream of a World-Class Society.* Cambridge, MA: MIT Press.

Wallstein, Bjorn. 2013. *Underneath Norrkoping: An Urban Mine of Hibernating Infrastructure, Licentiate Thesis*. Linkoping: Linkoping University, Sweden.
Wharton, Amy S. 2009. The sociology of emotional labor. *Annual Review of Sociology* 35: 147–165.
White, Richard. 1995. *The Organic Machine*. New York: Hill and Wang.
Winddance Twine, France, and Jonathan W. Warren (eds.). 2000. *Racing Research, Researching Race: Methodological Dilemmas in Criticial Race Studies*. New York: New York University Press.
Winner, Langdon. 1980. Do artifacts have politics? *Daedalus* 109: 121–136.
Wolf, Margery. 1992. *A Thrice-Told Tale: Feminism, Postmodernism, and Ethnographic Responsibility*. Stanford, CA: Stanford University Press.
Zuboff, Shoshana. 1988. *In the Age of the Smart Machine: The Future of Work and Power*. New York: HarperCollins.

10

Inside the Bicycle: Repair Knowledge for All

Tim Dant

Introduction

Contemporary capitalist society has been built on the specialised manufacture of a wide range of domestic and personal consumer devices that have either been designed not to be repaired, or to be taken to specialist workshops for repair, often those operating under the approval of the manufacturers. The increasingly complex technology employed in these goods apparently does not lend itself to do-it-yourself (DIY) repair but users—and sometimes manufacturers and professional repair agencies—have changed the potential for DIY repair with online videos that are easy to make and free to view by anyone who has access to the Internet. This paper will explore how these videos work by taking an ethnographic approach to understanding what they show and how their users might interpret them. The importance of repair videos lies in their potential to encourage repair and conserve useful objects and materials, countering the effects of contemporary design and manufacture,

T. Dant (✉)
Lancaster University, Lancaster, UK

I. Strebel et al. (eds.), *Repair Work Ethnographies*,
https://doi.org/10.1007/978-981-13-2110-8_10

which promote a throwaway culture that uses energy and depletes natural resources. Industrialisation, standardisation and mechanisation have led to the removal of manufacture from the context of use and made things less amenable to repair and maintenance. Can online repair videos empower users of things to possess them more fully by understanding their inner workings and taking responsibility for their repair and maintenance? After some background on DIY repair, this chapter will explore two examples of online videos, firstly replacing the battery in an iPod and then at more length, the repair and maintenance of the bearings in a bicycle wheel.

DIY Repair

Hans Jonas's revision of Kant's categorical imperative gives a new moral injunction that underlies being human: 'Act so that the effects of your action are compatible with the permanence of genuine human life' (1984: 11). Jonas argues that our first responsibility is to other human beings, including those not yet born, but this involves a changing moral relationship with technological objects, things made by humans that have consequences for the lives of others. By conserving objects we can reduce the consumption of scarce resources and energy and accept what Jonas (1984) calls the 'imperative of responsibility' for technology. One way that we as users of the material world can take responsibility for it is by learning how to repair the things we have (Henke 2000; Graham and Thrift 2007; Gregson et al. 2009). In doing so, we learn about how they are made, what materials they use and how their useful life can be extended to save the energy and raw materials needed to replace them. This paper discusses how new technology—cheap to make and free to access short videos available on the World Wide Web—can enhance the capacity of everyone to repair existing technologies.

The idea of 'do-it-yourself' (DIY) emerged towards the end of the nineteenth century linked to the extension of home ownership and working as an employee rather than for oneself (Gelber 1997). It was in the 1950s that it became established as a mode of domestic work that applied particularly to decorating and fixing things around the home.[1] DIY usually refers to work on the shaping and structure of the domestic environment that became the focus of the consumer culture of shops, magazines, manuals, commodities, advertisements and television programmes during the last half of the twentieth century (Watson and Shove 2008). DIY also extends to mechanical work on cars, bicycles and boats where the difficulty and cost of repair often involves getting the repairer to the machine or the machine to the repairer. There are also more 'niche' DIY cultures such as those around electronics and digital equipment with manuals and magazines as well as support groups. The World Wide Web has provided a new portal for sharing non-professional knowledge of all sorts of technical know-how, allowing interests in hobby making and DIY to be shared regardless of location. As Leadbeater and Miller (2004) point out, the Linux operating system is the creation of a widely distributed 'pro-am' group with sufficient knowledge of digital systems to devise and modify what has amounted to a serious challenge to professionally generated commercial computing systems. The participants are amateur in the sense that they earn nothing from their participation but professional in the standard of knowledge they draw on to contribute. The contemporary pro-am culture is often enabled by information technology as well as contributing to it (Leadbeater and Miller 2004: 41).

The act of repairing things changes human relationships with the material world. At the point of purchase consumers are little interested in how a thing is manufactured, its inner workings or even the materials it is made of, whether it is a music player, a pair of trousers or a bicycle. Instead they are interested in the object in itself—its efficacy, affordability and aesthetics. They ask of it questions that orient it to their wants; does it do what I want at a price I am willing to pay and still meet my aesthetic standards? But once acquired and taken up into use, the objects and equipment of material life 'withdraw in order to be ready-to-hand' (Heidegger 1962: 99). They become an extension of the actions of our body to achieve what we want in the world. The music player is used for playing music, trousers are simply worn and the bicycle ridden. The focus of attention shifts from the thing itself to what it does, what it is directed towards and what it is useful for. Heidegger tells us that ready-to-hand is the normal mode of being for things but there are circumstances when a thing, a material entity, becomes a matter of concern such as when they are '…not properly adapted for the use we have decided upon. The tool turns out to be damaged, or the material unsuitable' (Heidegger 1962: 102). When a thing, or piece of equipment is demonstrably unready-to-hand, when it does not work as we expect, it becomes conspicuous: 'Pure presence-at-hand announces itself in such equipment, but only to withdraw to the readiness-to-hand of something with which one concerns oneself—that is to say, the sort of thing we find when we put it back into repair' (Heidegger 1962: 103). The thing that no longer works '… reveals itself as something just present-at-hand and no more, which cannot be budged without the thing that is missing. The helpless way in which we stand before it is a deficient mode of concern, and as such it uncovers the Being-just-present-at-hand-and-no-more of something ready-to-hand' (Heidegger 1962: 103).

When something fails to work as we wish, we take notice of it and give it attention. We turn our attention to the music player that seems 'dead', the trousers that are torn or the bicycle that makes a horrible noise. Our helplessness at the thing's failure may lead us to discard the object and get another one, or we may pass it over to someone else for attention and repair. But if we embark on a repair ourselves we engage with the thing in all its complexity and get to know it in a different way; we get inside it, beyond the veil of its outward appearance that makes a smooth unitary thing of all its inner workings. As we become more intimate with it as an object with an interior and invest our labour in it, we make it 'ours' in a different way than when we just buy and use it. We have to get inside the music player, understand the seams and stitching of the trousers, and distinguish the moving parts of the bicycle. Our eyes and our fingers touch the parts of the object that we've not seen before as we get to know the thing; the materials it is made of and their properties, the interaction between parts and their workings, the way it was made.

Repair Videos—The iPod Battery

Replacing the battery in an iPod is a good example to illustrate how repair videos can enhance the responsibility of users for their things. Traditionally batteries are easily replaceable, usually disposable, energy sources for portable electrical and electronic devices that have compartments with doors and contacts for the user to fit them. Replacement batteries—even those designed to be rechargeable—are available in a range of standardised sizes and variable strengths but all have a limited life. However, a new generation of devices has incorporated internal rechargeable batteries to which the user is not intended to get access; the device must either be thrown away or taken to a specialist service when the battery fails to recharge, which is typically long before the device itself fails. Apple Inc. are famous for creating consumer equipment with sophisticated digital electronics embedded in objects with distinctive naturalistic shapes that shroud the structure, mechanics and electrical and electronic components from view. Their portable devices have taken advantage of the reliability and efficiency of embedded internalised batteries and the user is not intended to replace them or undertake any other repairs: Apple offers service through a carefully controlled network of centres. However websites including WikiHow, HowCast and iFixit as well as a number of amateur repairers confront this corporate control of repair head on by providing online instructions on how to, among many other things, replace iPod batteries and hard disks. Some use video to show the physical techniques and the use of tools that need to be applied to repair these objects.

At the time of writing it is possible to send your Classic iPod to Apple in the UK to get the battery replaced and it will cost £52.44 plus the cost of posting and packing it. In the UK a battery and fixing kit can be bought for £12.93. Although many domestic objects that use batteries (torches, clocks, travel razors) are designed to be opened and the batteries replaced, the rechargeable batteries used in devices including the iPod and some other small equipment (e.g. rechargeable razors and toothbrushes) are soldered into the structure behind casing which is not designed to be opened. When it was first introduced in 2001 the iPod was distinctive in lacking knobs or switches or an access 'door', cap or hatch or even screws by which to get at compartments or components. The design distinguished the iPod from the Walkman style cassette players that had been used as personal music players for the previous decade. The Walkman took standard disposable batteries, had press buttons, a knurled volume wheel and opened to enable cassettes to be slotted in. In contrast the iPod had a single flat scroll wheel (originally mechanical, latterly electronic) with electronic control buttons slightly raised from the surface (later incorporated into the scroll wheel). The effect was to create a 'pebble' shape, a design form to fit in the palm of a hand and slide without hindrance into a pocket or bag.

A two-minute professionally crafted HowCast video[2] available on YouTube, shows how to change the battery on an iPod Classic with a slick presentational style reminiscent of advertisements. It uses electronic music and a distinctively American, female voice who begins by repeating the subtitle 'you can pay Apple to replace your iPod battery or you can save big bucks by doing it yourself'. There are only 9 steps in the procedure, the first being to buy a new battery online.[3] Step 2 shows video viewers how to use a flat piece of durable plastic ('a guitar pick works well') to pry open the casing by poking inwards at the tabs holding it in place. The voice-over follows the script of the written advice but the video shows head and shoulders images of the process and some close ups of hands and tools working on the object.

The video uses stop motion sequences showing objects moving on their own, has text overlaying images, magnification to show details, tips on working procedure and numbered steps. The work is 'shown' to the camera and video viewer as a demonstration to an apprentice might be, with tightly edited transitions between head and shoulders views of the repairer and close-up shots of the objects. The repeating rhythmic music and the matter-of-fact woman's voice-over make light of what is visibly a tricky task. The editing of shots around the 'steps' turns the flow of repair work into a systematic procedure making it appear more 'doable' to the novice. The positioning of hands and tools in the video direct the viewer's attention to the different parts that need to be detached—hard drive, battery—and then reattached. Each 'step' can be taken as a task in itself and the work can pause once each is completed. However, editing into steps and short scenes reduces the detail, both in terms of verbally delivered advice and in terms of seeing the gestures of the work unfold. Although six years old the HowCast video shows 1,047,658 views and 998 'likes'.

In contrast, a more leisurely video by 'harrymedan'[4] of the same task on a 5th Generation iPod Classic takes 10 minutes 46 viewing time and shows the task from the perspective of the repairer. This amateur video has no music, poor lighting and focus and an opened cardboard box acts as a backdrop. The male voice speaks leisurely and we can see the hands actually begin and achieve the movements of opening the casing. Some procedures—such as fixing the ribbon connector—don't work first time and need to be repeated. Segments of the video follow the work in the real time it took to achieve and we hear the commentator speaking as he works. At times you can hear cars honking in the street outside and the repairman's strained breathing when there is no talk. There are edit breaks where the camera position is changed; from looking over the repairman's shoulder, to looking onto his hands as if from his chest. This video has 114,560 views and 486 likes.

There are a number of amateur and professional videos, some linked to selling spare parts, that show the same task on different generations of iPod Classic. The procedure appears to get more difficult with successive generations of iPod as soldered instead of pushfit plastic connectors, overlaying parts and an increasing number of metal casing clips (13 on the 6th Generation) are introduced. No doubt these changes to design, which reduce the repairability of the device, were a corollary of packing more functionality into a similar sized casing. One ten-minute video by 'quickconnection'[5] even shows the damage done to the edge of the metal casing by attempts to open a 6th Generation iPod Classic. Surprisingly this video had been viewed 77,580 times on YouTube and many of the comments said how helpful it was … although there were only 85 likes. The iFixit site[6] classes the task of opening the 6th Generation iPod Classic as 'very difficult' and presents it in 24 steps using high-definition still photographs from the perspective of the repairer; the first 16 steps just show the technique for opening the casing!

Sealed units, and beautifully encased electronic devices like the Apple iPod have pretended that their workings were magical; continually available, beyond inspection, never needing servicing or maintenance in their lifetimes. The use of specialist, often tiny, 'tamper proof' screws, hidden clips and catches, item-specific tools and fine torque wrench settings obstruct DIY repair. But users' experience tells them that manufactured objects do not simply 'die'; unless their shape and structure has been completely destroyed, their failure to function is usually linked to the failure of a component responsible for a mechanical, motor power, electrical, electronic or digital function. And the Internet has provided resources for resisting throwing away consumer objects, for resisting the conceit of manufacturers that they know best, that their 'designated repair' centres are the only appropriate sources of repair services.

Repair Videos

Anyone with access to the Word Wide Web via a computer, smartphone or electronic tablet can view videos and sources of advice freely in the public domain as well as buy specialist tools and replacement parts for repairs. Advice is sometimes provided through generic sites such as HowCast, WikiHow,[7] iFixit[8] and Instructables[9] that gather articles written by volunteers to show how to do things in a series of steps, often with colour photographs. The sites sometimes include 'tips', downloadable pdf files of the steps, or videos with commentaries that cover the steps and may also present repair as an ethical and political issue.[10] Other free advice about repair is available through online 'forums' where, for example, the owners of a particular make and model of car will share tips on how to DIY repair particular faults.

Some videos are professionally made and linked to commercial services but many are made by enthusiasts using domestic video technology. Unlike advice videos on VHS, printed magazines or repair manuals published and sold commercially, most online repair and how-to videos advice are free at the point of use. Most videos are short—less than two minutes—but even longer ones rarely stretch to fifteen minutes. The videos are articulated and presented in terms of a repair task that might confront a user rather than sections that build into a course of systematic knowledge. This practical and situated knowledge can be accessible alongside and during the attempt by someone to undertake a repair. The videos can be stopped, replayed or taken step by step, with the repairer listening and watching the step and then undertaking it before moving to the next one on the video. Repair manuals are usually arranged in a systematic way, organised in sections with a contents page and an index by which the repairer finds the section they need to tackle a task. In contrast, the videos are most easily found by searching the whole internet with some keywords; the search will return web page addresses and short introductions, including those for videos, and the repairer can choose those most relevant or they can repeat and refine the search with different keywords. The use of repair manuals is most likely to be occasioned by the need to undertake a repair task, and the same is the case of repair videos—neither are likely to be read for pleasure or for general learning. However, the repair manual is a structured text that offers one approach to a task made available through an author and publisher. Using the Internet to access repair videos gives the repairer the opportunity to view a number of different approaches and choose which ones to make use of.

Enthusiasts provide the core content for these websites without payment but the sites still need to generate enough income to pay for a core staff to run the site and maintain its quality and identity. Revenue is raised by advertisements and sometimes sponsorship[11] and some sites such as How to Repair Home Appliances[12] and Espares[13] are run as commercial companies that offer how-to guides linked to spare parts they are selling. The origins of these websites are often in the digital technology culture of the west coast of the United States and offer a particular expertise in the repair of electronic devices. An aspect of the repair ethic that some of the sites follow is one of resistance to the major digital electronic companies that sell attractive, expensive devices but offer little advice about how to repair or maintain them.

In their presentation of occasioned, situated, linear, procedural and practical knowledge as opposed to abstract, systematic and theoretically organised knowledge, repair videos are noticeable for providing instruction in those types of repair that address the structural form of the object. From the perspective of repair, the history of technological progress can be thought of in terms of seven modes of creating things for human use. The first mode is about *shape* as material forms are created that are useful for everyday life; wood, bone, skin, stone, metals and clays are shaped into tools, knives, spears, clothes, pots and so forth. Attaching components to one another to make an object leads to the second mode, that of *structure* as shapes are built to bear the weight of other things (including people), to produce furniture, buildings, boats, sledges and so forth. The third mode is that of *mechanics* in which wheels, pulleys, levers and gears are used to create objects that adapt wind or animal power to achieve movement in objects; sailing vessels, mills, cranes, carriages and so on. The fourth mode is the introduction of *motor power*—engines driven by steam, petrol and diesel—that combine with mechanics to produce machines with an independent source of energy to generate movement. The fifth mode is the use of *electricity* as an adaptable and storable source of energy for lighting, heating and motor power in a variety of structures and machines. The sixth mode is the use of *electronics* that uses electricity to transmit signals and messages between people but also within things from sensors to indicators and dials, from controls to servo systems and so on. The seventh mode is that of *digital computing* in which the signals of electronics are linked to memory that can store algorithms or programmes to control and manage subsystems.

Each mode involves a new transformation of materials found on earth and a further adaptation of the physical properties of things to realise ends that are useful for human existence. Each mode requires new skills and tools of manufacture that become progressively more oriented to specific and abstract knowledge introducing new complexities of repair. What is striking about the sorts of repairs tackled in repair videos is that they usually address a failure in the **structural** relationship of components. If an individual component has failed—e.g. the battery in an electronic device—the repair is about removing that component from the structure and replacing it with a working component. In those structures that have a mechanical dimension there is an added dimension of cleaning and lubricating that may be considered as a maintenance task or be necessary to repair the item and return it to full working order. Even in a device that uses electricity, electronics and digital computing such as an iPod (each of which mode requires different levels of abstract and systematic knowledge to fully understand), the topic of most repair videos addresses the structure of components. Videos on repairing an electric iron for example, a technology from the fifth mode that was first developed at the end of the nineteenth century, will focus on replacing parts such as the fuse, the cable or the thermostat. Repair seldom involves remodelling a component or forming it from scratch, instead a replacement is inserted into the structure. Most DIY repair videos provide knowledge of what is going on inside an object by revealing the structure and mechanical workings that are obscured by casings and other components in normal use.

Repair videos are often occasioned by the simplicity of the knowledge needed to return an item to full function. Diagrams including exploded diagrams in still images—either on paper or on a web page—can show structural relationships between components but video can show both how moving components interact and the sequence of gestures of hands and tools that is needed to work on the structure. As a number of commentators have pointed out, photorealistic imagery presents a 'superabundance' of information, much more than is included in narrative or even descriptive exposition (e.g., Sartre 2004: 18; Ellis 2002: 19). But film and video offer this superabundance of information in a 'continuous present' in which the temporality of what happens on the screen is the same as for anyone who is actually present for those events, representing what is going on in a degree of fullness not available in other media (Dant 2012: 109–111). The realism of the video footage combined with the continuous temporality and the coterminous ambient sound, means that the detailed process of a repair can be shown and the insides of an object can be literally revealed. Repair videos embody the work of repair, situating an identifiably real person and their body in relation to the object, components and tools. Professionally prepared videos often use a voice-over commentary reading from a prepared script but amateur videos typically use the voice of the person whose hands are visibly undertaking the task as they speak. The ease or difficulty, the degree of effort required and the precise orientation of fingers for gestures and movements is expressed in the body and voice of the presenter. The sound of objects and the voice of the person confirm the verisimilitude of digital video in its unfolding continuous present. The person's voice and body gives the work personality, the quality of a first person narrative that is different from the second person instructional voice of the repair manual with its outline diagrams or still images.

Inside a Bicycle Wheel

To explore how online videos work in more detail I have chosen to look at four videos that show how, through interacting with components, an amateur can undertake the repair and maintenance of the bearings on a bicycle wheel. Bicycle wheel bearings seldom collapse or lock up completely but they do deteriorate without regular maintenance to the extent of making a bicycle noisy and hard to ride. Disassembly, re-greasing and reassembly is both a 'maintenance' and a 'repair' task that requires getting inside the bearing and replacing any significantly worn or damaged components. The four videos have been chosen from amongst the many that appear on YouTube and are easily findable with a Google search for videos on 'bicycle rear wheel bearings'. The search produced 131,000 hits but included videos about motorcycle bearings and other repair and maintenance tasks in which wheel bearings were mentioned. The videos chosen were from the first thirty listed on Google at the time the research was undertaken and all show the repair and maintenance of a bicycle wheel bearing and all show the tasks of disassembly, cleaning, inspection and reassembly—each video takes the viewer inside the bearing at the centre of a modern bicycle wheel. Each was selected to show a range of different features in video repair and I have labelled the repairer and their video as 'Floorworker', 'Professional', 'Dirty Biker' and 'Music Maker' to reflect their different styles. Two videos (Professional and Music Maker) have been professionally made with commercial sponsorship and feature professional bicycle repair technicians. In one case (Professional) the video takes the form of a tutorial by the technician to a group of students, the other (Music Maker) has no commentary or 'advice' and simply shows from the perspective of the repairer, the process of disassembly, cleaning, re-greasing and reassembling a particular type of bearing used on some mountain bikes. Whereas in Professional the repairer is shown and speaks to camera, in Music Maker, the repairer appears only as a pair of hands working on the wheel. The other two videos (Floorworker and Dirty Biker) are apparently both made by amateurs who appear in the video and their voice-over commentary is in the same continuous present as their actions and gestures in relation to the repair. However, Dirty Biker's video is more complex with segments, including one presented as a tutorial to camera with diagrams and gestures to aid explanation. Other sections show the work proceeding and the view presented to the camera is close to the point of view of the repairer. Floorworker uses a static camera on a tripod that shows the presenter from the side as he works and describes what he is doing. The four videos contrast professional and amateur approaches, presentational styles and the use of the camera both for a tutorial format and to show the work from the point of view of the repairer.

The video analysis was undertaken by reviewing a number of videos that appeared to show the repair of the rear wheel bearing on a bicycle, which led to selecting four for closer analysis because they showed essentially the same task but in different ways. The analysis proceeded by viewing and reviewing each video many times, looking for points of similarity and difference. Video material is available for replaying either as a whole or in small sections so it is possible to review particular gestures or actions and compare them between videos. There are three aspects to the analysis: the performance of the repair, the presentation as instruction and the video technique (camera angles, close-ups, editing). A description of each video in note form was prepared with verbatim transcription of key speech items and many still images captured to identify points of contrast. Each video will be introduced in extreme summary before a discussion of some of the similarities and differences between them but first it is helpful to explain what they all show.

The bearings on a bicycle are mechanical moving parts that are normally hidden from view. Ball bearings were patented by a Parisian bicycle mechanic, Jules Pierre Suriray in 1869 and improved the smoothness, efficiency and comfort of the 'ordinary' bicycle. They rapidly become ubiquitous in bicycle technology so that all contemporary bicycles run on wheels with bearings and also use bearing races in a number of other components where metal moves inside metal (e.g. between the axle connecting the pedal cranks and the bottom bracket of the frame; between pedals and their axles; between the steerer tube and head tube at the headset and fork crown).[14] Ball bearings are largely invisible from the outside of the bike because they need to be enclosed in a race of some sort and they are often obscured by grease that is opaque and usually grey or black. To function efficiently as a bearing the balls need to be held together closely with the components that they are connecting and looseness or 'play' will undermine their smooth movement. On the other hand, being held too tightly will lead to friction and again restrict smooth movement. A film of grease between the balls and the race provided by the cupped interface with the components helps maximise smooth movement and minimise friction. The steel balls in a wheel bearing declare their presence, their function and their arrangement when they are disassembled. Disassembly gives the repairer a 'representational state' of the components; how many balls there are, what size they are, whether there is any grease present and whether the surface of the steel balls or the race is worn or scarred. However, in disassembling the axle to expose the bearings there is a risk that balls will escape and fall to the floor or workbench—alternatively some can be held in place by grease and need to be removed with a tool.[15]

Going Inside a Wheel Bearing

'Floorworker' (13:42)[16]

The first ranked video is a more recent and refined version uploaded by the same repair worker, 'RJ the Bike Guy', but this earlier video of his demonstrates some of the difficulties of amateur repair videos. Floorworker works kneeling or sitting on a floor covered with an old curtain that becomes the background to the scene with the camera looking down from a tripod on his right-hand side. Although not broken into stages there are a sequence of sub-tasks as the parts are removed and tools are deployed in disassembly of the bearing. Floorworker introduces tools and parts as they appear in the flow of work by holding them up for the camera and naming them (e.g. 'cone spanner', 'cone nut'). Tools are usually picked up from an area of the workspace to his right, at the bottom of the screen, where they remain literally ready-to-hand and in easy sight although not systematically arranged.

There is some editing with jump cuts and speeded-up sequences (e.g. of repeated gestures undoing nuts) but the task appears to have been captured in one continuous shot. The fixed camera angle means that some tasks are either too distant to be seen or they simply happen off camera. Some of the larger components are visible but smaller ones such as the cone nuts are too far away from the camera to be discernible on the screen. Although the smaller bits from the axle are laid out to show the sequence for reassembly they are then all dropped together into a can of 'paint thinner' to clean off the grease. As he takes off the cone Floorworker holds it up and says '…its got a little race around there for the bearings' but the cone is too small for the curve of the race to be visible on the screen and because the fixed camera doesn't follow the object, he actually holds it up outside the frame of the video.

Having cleaned the race with a paper towel and paint thinner Floorworker says 'OK, now you want to examine the races in here [gestures with finger] where the bearings go, for any cracks or damage … I don't see any. OK'. During reassembly of the cleaned and dried components, the tightening of the first cone against the lock nut is done slightly out of the top of the frame as Floorworker rises on his knees to bring the work closer while using both hands. Once he has greased and replaced the balls in the cup race in the wheel hub, he holds it up to the camera to show how they are located and with a finger gesture shows how they go round in a ring (Fig. 10.1).

With all the balls in the races and dustcaps replaced, Floorworker puts in the axle and adds the remaining cone, spacers and locknut. He then shows how the axle should turn freely without binding or play by

Fig. 10.1 Floorworker showing the balls in the bearing race (© Video still: RJ The Bike Guy, available at https://www.youtube.com/watch?v=IrzIHjhckSU)

holding the axle ends between his fingers with the wheel vertical from the floor and then moves them up and down to wobble the wheel saying: 'But you also don't want play in there, so you kind of move it around… you want to turn it nice and smoothly but not have any play in there'. Once happy with it, he shows how to use a cone spanner to hold the cone while using a crescent spanner to 'tighten the locknut into the cone'. Floorworker tells us to check, having done the tightening, that there is free movement and no play—he finds 'just a little bit of play' and loosens the locknut, then adjusts and retightens the cone and locknut.

Although we never see the repairman's face we hear his voice telling us what he is doing as the video progresses. His commentary was spoken while the video was shot rather than added later so the hands and body of the repairman make gestures that work with the voice to draw our attention to specific features of the work as it unfolds. He is not working from a script and his expression doesn't sound carefully prepared or rehearsed but is spontaneously spoken with hesitations, pauses and reformulations as the task proceeds.

Dirty Biker (16:17)[17]

Dirty Biker uses a number of fixed camera angles and begins with a shot of himself and the bicycle as he introduces the video. Unlike Floorworker's video which more or less follows the sequence and temporality of the task, Dirty Biker's video has a number of cutaway sequences inserted. In one he gives a lecture-style presentation to explain how the ball bearings work using coloured diagrams, a physical hub, words and hand gestures to point to features he is describing and a whirring gesture with his fingers to indicate the rotation of the bearings (Fig. 10.2).

From the beginning Dirty Biker explains how the bearings work to reduce friction and shows you on his well-used hybrid bike that has mudguards and a plastic crate on the rear rack. There are rust spots and dirt on the bike and his hands and fingers have dirt and grease on them from working on the bike.

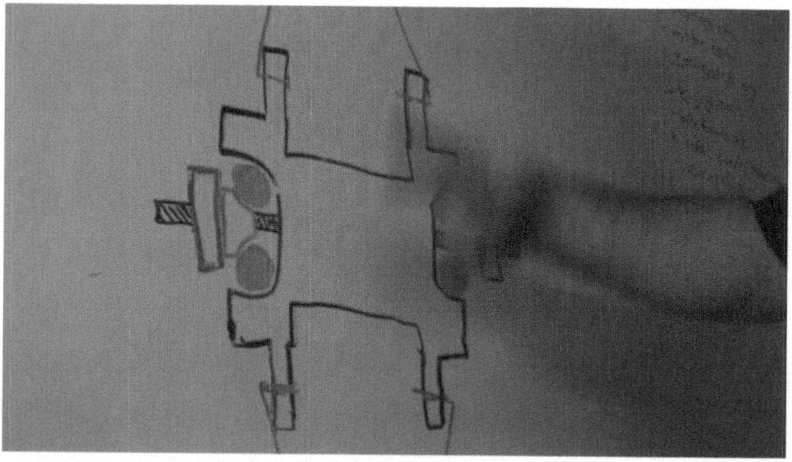

Fig. 10.2 Dirty Biker's diagram and rotating gesture (© Video still: C. Hoyle, available at https://www.youtube.com/watch?v=LNzN9oTQA8s)

Dirty biker uses close-ups to show the use of the cone wrench and adjustable wrench to unlock cone and locknut by twisting in opposite directions. He has cut a convenient hole in his wooden bench so that, with the wheel on its side, one end of the axle projects down through the bench and he can hold it from underneath as he undoes nuts on the top end of the axle. As the cone nut is removed he brings it close to the handheld camera, but it stays in focus, so video viewers can see the curved shape that makes up half of the ball race. Then the camera moves down towards the bench and allows viewers to see the ball bearings in the cup half of the race inside the wheel hub (Fig. 10.3).

Once the components are removed and cleaned using a rag, he shows them in tight close-up, effectively magnifying the view of their surfaces, and remarks on the pitting on the cones which he replaces with a clean, although used, pair. He explains that when the correct number of balls are replaced in the race… 'It looks like there's a bit of a gap in between each ball bearing, and that's what you want, I'm pretty sure ten is the correct number in here [it is actually a front bearing], um, you definitely don't want them so that they're so tight in there that they're always touching each other um because they are, each ball bearing is

Fig. 10.3 Dirty Biker showing cone and bearings (© Video still: C. Hoyle, available at https://www.youtube.com/watch?v=LNzN9oTQA8s)

rotating on its own, so you don't want them to be rubbing against one another too much'.

In the reassembly Dirty Biker repeats and demonstrates with the reassembled axle what he explained in the abstract with the diagram about the bearing being tight enough but not too tight explaining that 'this step requires quite a bit of trial and error' because as the locknut is tightened it will shift the cone. At a few points during the video, Dirty Biker has overlaid labelled buttons that allow you to click through to his other tutorial videos e.g. for removing the rear wheel cassette. He ends the video with his face to camera through the wheels of his upturned bike, thanking people for watching and asking them to subscribe.

'Professional' (15:14)[18]

This video contrasts in style from the others because it is shot with two handheld cameras not controlled or directed by the Professional repairman who is teaching three students how to undertake the service of a bicycle rear hub (they are shown in a brief cutaway shot standing in front of him). The students get a front-on view as Professional addresses them over the wheel he is working on while the videoview is from the side and over his arms. Professional wears a uniform with 'ATG Training'[19] on the back and different logos on his work overall and gloves and he stands in front of a metal workbench with tools arranged and fixed to a pin board at the back. His voice is measured, steady and presents the task slowly and in stages with comments about technique. He says: 'There's a lot of parts to consider so the layout really is paramount. I try to work to the left and the right of the bicycle, so we know the gears are on the right-hand side, so that's mirrored in my layout here.' Professional bends over the wheel from an upright position to remove the freehub and loosen the rotor disc for the brakes but when it comes to removing the nuts from the axle he turns to one side and secures the wheel horizontally in a bench vice.

As Professional cleans the old grease off the bearing race he says: 'What I'm able to see now, is any, what they call pitting or brindling' (Fig. 10.4).

Grease is removed using workshop absorbent paper taken from a roll on the end of the bench and Professional tells us that he applies some 'rapid degreaser' to help with the cleaning. But he doesn't show the clean race to the camera and he doesn't explain what you might do should you see such pitting or brindling. Having put fresh grease into

Fig. 10.4 Professional showing the cup (Online video: available at https://www.youtube.com/watch?v=yHvpzlcKBKg)

the clean race he shows how to put in the ball bearings by holding the wheel in his left hand at waist height at an angle of about 45 degrees so that with his right he can put each ball into the grease.

As he pushes each ball bearing into place he says: 'There always seems to be space for one more ball bearing. Don't be tempted to go and get a slightly smaller ball bearing and drop it in.' While he explains that this would damage the ball bearings he doesn't either explain or show his audience how the bearing actually works. Once the axle is reinserted the wheel is returned to a horizontal position in the vice and as Professional replaces the second cone nut he says: 'you are just looking for this to be finger tight … just to remove the play from the bearings, it doesn't need to be excruciatingly tight otherwise your axle won't turn'. He says that he tightens it until he can feel the pressure on his fingers and then 'I back it off a quarter of a turn'. Professional does explain, pointing to the components, how turning the lock nut can tighten the cone nut so they have to be tightened against each other with each held by a spanner.

Once the bearing is reassembled with the axle in place and the nuts tight it should be checked and Professional shows this by holding one end of the axle and moving it up and down so the wheel rim moves from side to side. He says: 'I'm wobbling the axle there to see if there's any play. Now if you can feel the sort of play that you're not sure if it's the bones in your hand that are moving, or whether there's play in the axle, then that's OK because when you do up the quick release you'll actually tighten the hub together'. What is striking about the Professional video is that we never see the bearings in place in the wheel hub and the axle is taken out at a distance from the camera.

Music Maker (9:50)[20]

This video, showing the servicing of Shimano XT mountain bike bearings, has continuous electronic music, no talk and just one text frame at the beginning explaining that the video shows a 'highly experienced and qualified mechanic' and 'special tools'. The repairer is simply doing the job and is making no attempt either to teach students or those watching the video; he does not pause or get distracted by having to explain what he is doing and we never get to see more of him than his arms or hands. The camera work is mainly handheld but is steady and usually is shot over the right-hand elbow of the repairer with very good close-ups. The editing includes fades from one shot to another that provide a sequence of scenes as well as skipping bits in which not much happens; nonetheless there is no articulation of steps, stages or tasks. We watch as lock ring, brake disc, locknut and cone are removed from the wheel that is horizontal on a well-lit, metal bench (Fig. 10.5).

The wheel is then turned to be upright on the bench as the axle is withdrawn—unlike the axles in the other videos, it is a much larger diameter hollow anodised aluminium tube smeared with grease. We see the axle and balls cleaned with the same type of blue workshop paper that Professional used and then the cup race is cleaned with degreaser. Before reassembling the axle, with a slow panning shot the camera shows us the balls arranged in the cup race in the wheel hub that, unlike the bearings in the other videos, incorporates a partial cage to hold the balls in place (Fig. 10.6).

The balls are carefully replaced with the wheel upright on the bench, the grease, applied with a grease gun, holds them in position until the axle is reinserted. The cone is replaced and we see the hands testing it for movement up and down and rotation before the lock ring is replaced and then checking again for ease of rotation once the lock ring

Fig. 10.5 Music Maker using cone wrenches (Online video: available at https://www.youtube.com/watch?v=wWEoGm1XY7U)

Fig. 10.6 Music Maker replacing balls (Online video: available at https://www.youtube.com/watch?v=wWEoGm1XY7U)

is tightened. The last stage of assembly involves returning the mounting for the disc brake rotor, the disc itself and then its lock ring for which a specialised tool (a large spanner with notches) is needed.

Discussion

The four videos are showing much the same task and each takes the viewer 'inside' the wheel bearings on a bicycle but they do it in different ways. There are some differences to the task as Professional and Music Maker work on modern mountain bike wheels with disc brakes and Dirty Biker and Floorworker show us the task on traditional road bikes. Dirty Biker and Music Maker actually show the disassembly of front wheels whereas Professional and Floorworker show the disassembly of rear wheels with cassettes and spacers.

Each repair worker has a 'phenomenal field' (Dant 2005: 93), a space of action for the repair work that is oriented to the worker's body, particularly their eyes and hands. Tools are usually on the periphery of the field and are brought into action by the hands. In use, tools and components are brought close enough to the front of the body by the hands so that whatever is being worked on is roughly equidistant from each eye and each arm. The work is ideally done with the object at a distance from the abdomen such that the arms are bent at the elbows and the hands, wrists and forearms are free to manoeuvre in their full range of movement and force. The phenomenal field is construed by the posture and orientation of the particular repair worker and varies between and during the videos—it may show a bias towards the right hand which tends to exert force and use tools while the left hand holds tools and objects. Professional stands facing his students and works mainly holding the wheel in the air in front of him, often tipped at an angle of about 30 degrees towards his students. When he turns aside to use the bench vice to hold the wheel he unusually works at an oblique angle to the objects to keep the left side of his phenomenal field open to the view of the students (even though this sometimes means he obscures the videoview of the work). Music Maker stands and works at a bench throughout with the wheel at times horizontal and at other times vertical but always in front of him. Dirty Biker sometimes faces the camera over or beside the wheels on the bicycle and sometimes works at a bench with the wheel horizontal on it. Floorworker is unusual in working sitting or squatting on the floor, largely with the wheel in a vertical plain but sometimes held by his left hand at an angle of about 45 degrees tipped obliquely towards the camera to give his right hand maximum control.

The orientation of the camera to the phenomenal field varies between the videos giving different points of view (POV) of the work and the components. For Professional, the two handheld video cameras are on either side of his phenomenal field while his students are opposite him. Dirty Biker gestures to the diagrams or to the wheels on the upturned bike with the videoview over his left shoulder and turns to quarter face the camera as he talks. At other times he arranges the camera to take a close-up from the left-hand side of his phenomenal field and because his left hand is often out of sight this view is almost that of the repairer. Then to show the cone nut and the bearings in the cup race he holds the camera in front of his face giving viewers a repairer's POV. Music maker's video alternately gives an onlooker's view over the repairer's right elbow and then gives almost the POV of the repair worker by looking over his right shoulder in close up. Floorworker has set up a tripod so the screen view is always 90 degrees to his own with the viewer looking on from the side at about 140 cm from the ground. The camera does not zoom or move and loses focus when things are held up to it.

A particularly difficult stage in this repair task is reassembly. All the parts need to be replaced in the correct order and two workers (Floorworker and Professional) show how laying out spacers, cones and nuts with the axle can help. Working on a rear wheel is trickier because the threading of the axle is different as each side has different spacers. But the really tricky part of reassembly is in getting the cones to seat correctly on the ball bearings—neither too tight nor too loose—and video is able to show the particular gestures needed. First is the gesture of tightening the locknut with one spanner while holding the cone steady with a cone spanner. Second is checking for play and free movement in the bearing with the hand. Both Professional and Floorworker do this by holding the axle at both ends so the wheel can spin and then by rocking the wheel in the opposite plane with the rim waving from side to side. Both Dirty Biker and Music Maker hold a single end of the axle and move it in the bearing, firstly by turning it on its axis and then by wobbling it in the opposite plane. The repairer has to judge through feeling in the hand just when the bearing is right and then adjust it in a process of trial and error. Both Floorworker and Dirty Biker have to repeat the adjustment but Professional gets it right first time and so, apparently, does Music maker.

Conclusions

Many repair videos, including some made by amateurs, have an introductory sequence where tools or parts to be used are shown to the camera but this does not happen in any of this corpus of videos. Both Professional and Music Maker are clearly professional bicycle repair mechanics whose expertise is being captured on camera but Professional is teaching while Music Maker is simply doing the repair. Although Dirty Biker's workshop looks rough and ready and his bike is well used, he has a wide range of cone spanners and a confidence that suggest he has done this job many times on different bikes. His video is devised to show in-focus, close-up details of the parts and tools that have been used, sometimes giving a repairer's POV. He also gives instruction, sometimes with diagrams, in how bearings work. Professional is at ease with the instructional mode of his video and explains things fully but many details of actions and components are not shown in close-up and camera angles often don't give a clear view of the phenomenal field. Floorworker makes no claim to be professional and shows the action in a domestic setting rather than a workshop. The fixed camera is limiting in what it can show and the video provides demonstration with commentary rather than instruction. There is considerable variation between the different videos and the number of views and 'likes' gives some indication of how useful they are to viewers (see Table 10.1). Unlike using repair manuals, it is easy for viewers to watch a number of different videos of the same repair task (or the same video a number of times), taking different guidance from each viewing session.

These four videos show the potential of video to reveal the inside of mechanical objects with moving parts—something that is often difficult to imagine or explain in the abstract. The medium of video is able to show the bicycle user the balls and race inside the bearings on which they ride. The moving image in the video is able to show dynamically the hand gestures of undoing threads, removing and replacing the ball bearings and the fine adjustment of resetting the bearing that could not be shown in still images or diagrams. The material interaction between hands and things, between different materials and different orientations, is shown as real-time movement. The hardness of metals as against the tactility and mobility of grease and the variable resistance of screw threads are all demonstrated dynamically. Tools are shown in use and so are particular gestures of the working hand such as using cone spanner and wrench in concert by pulling them against each other to loosen or tighten a locknut; the degree of pressure used in the hands, the gradual pace of the gesture as it faces resistance are visible in the moving image. However, it is worth remarking that the videos cannot actually show the bearings moving under the weight of bicycle and rider.

Video has the capacity to show a 'continuous present' of the flow of movement and gesture that 'appresents' a reality and makes sense to the viewer through their experience of everyday life (Dant 2012: 109–116). This characteristic of the medium enables the viewer to experience

Table 10.1 How to… service wheel bearings

	Date	Videos made	Subscribers	Run time	Views	Likes	Dislikes	Comments	URL
Floorworker	26-4-2012	261	5232	13.42	131,548	174	21	65	https://www.youtube.com/watch?v=IrzlHjhckSU
Music Maker	31-10-2011	44	568	9.5	29,010	25	3	7	https://www.youtube.com/watch?v=wWEoG-m1XY7U
Professional	20-5-2011	2	203	15.14	140,491	316	14	53	https://www.youtube.com/watch?v=yHvpzlcKBKg
Dirty Biker	31-3-2012	44	568	16.18	208,525	1107	37	306	https://www.youtube.com/watch?v=LNzN9oTQA8s

something of what it would be like to be actually there, watching the action for real. Video reveals in segments of real-time events that are as if they are happening now, in the presence of the viewer. The medium requires no special skills of viewers, such as being able to read, and even the conventions of representation do not need to be understood or decoded as they do with an exploded diagram. Three of the videos include commentaries in English, but one is able to show the repair process without any spoken language—much could be learnt from the other videos even without understanding what is said.

When a user of something repairs it, they take responsibility for it, replacing what is missing and returning it to readiness-to-hand. We accept responsibility for the thing because as we confront just what is failing to function and we must consider just how it *should* fit into our life if it was ready-to-hand. Instead of simply expecting it to work in the way it was designed and originally made, we accept responsibility to help it to work in that way. Even if, finally, it has to be thrown away, we can do that knowingly too. The work of repair means getting to know the object better than we do as a user because we must learn to identify the causes of damage or wear and understand the consequences and how to deal with them. The science of materials or mechanics, the principles of electricity or electronics are abstract corpuses of knowledge that can be ignored as a user of technology. The repairer may also take a pragmatic attitude to learning only what is sufficient for the task in hand along with the use of special tools and safety issues—but she or he must at least understand how the object is *structured* and how its components contribute to its working.

Free to use and easy to access online videos promote the 'imperative of responsibility' by sharing knowledge about the repair of contemporary material culture and encouraging DIY engagement amongst users. The variations between the videos both in the way the repair task is demonstrated and the technical aspects of video preparation (camera angles, editing, quality of image, etc.) show that there is no single or 'correct' way to undertake the repair work or to represent it. Nonetheless, each video provides demonstration and explanation in a way that a live tutor might, to show consumers and users just how to undertake what might otherwise be daunting repair tasks. Repair videos may focus on the structure of material objects and the replacement of components rather than making or innovating a solution to the failure of a material device. They may not keep pace with the advancing systems of knowledge that come with newer modes of technology. But they nonetheless enhance the do-ability of repair by everyone and even just watching the videos increases users' responsibility for their devices and machines by simply knowing how they work.

Acknowledgements I am most grateful to Christopher Hoyle (www.cjhoyle.com) and RJ The Bike Guy (www.rjthebikeguy.com) who gave permission to print images taken from their online videos.

Notes

1. DIY: 'Tasks carried out within a household which could have been performed by a paid specialist. The term covers mainly decorating, home improvements, and motor repairs. DIY stores sell materials for use in such activities. DIY activities are encouraged by the tax system, which requires employed labour to be paid for out of taxed income, whereas family labour is tax-free' (Black et al. 2012).
2. https://www.youtube.com/watch?v=2h53otOBTsI.
3. http://www.howcast.com/videos/17489-How-to-Replace-Your-iPod-Battery. Although HowCast does not say so, it looks from the image like an iPod Classic 4th Generation. iFixit offer a 12 step procedure for the 4th Generation.
4. https://www.youtube.com/watch?v=HbNKyauCkh8.
5. https://www.youtube.com/watch?v=pLHY_BqhqKg.
6. https://www.ifixit.com/Guide/iPod+Classic+Battery+Replacement/561.
7. http://www.wikihow.com/.
8. https://www.ifixit.com/.
9. http://www.instructables.com/.
10. iFixit not only offers how to guides, it also has a manifesto (https://www.ifixit.com/Manifesto), runs an activist blog (http://ifixit.org/category/activism/) and an environmental page (https://www.ifixit.com/Info/Environment). WikiHow presents itself as a 'community of

knowledge philanthropists' who want to create a world 'where anyone can easily learn how to do anything' (http://www.wikihow.com/wikiHow:About-wikiHow). Instructables began with a group documenting individual projects for making things online: 'From cooking to 3D printing, to making just about anything fly, Instructables became the recipient of countless hours of tinkering, soldering, stitching, frying, and fun, making just about anything' (http://www.instructables.com/about/).

11. 'Instructables' collaborates with Intel to present a page about connecting things to the web—an 'Internet of things hub'—http://www.instructables.com/id/intel/.
12. http://www.how-to-repair.com/.
13. http://www.espares.co.uk/.
14. Ball bearings have been superseded on many more expensive modern bikes by roller bearings in sealed races.
15. It may be that the bearings are in a cage—which makes this part of the task much simpler—but this is not typical of good quality bicycle wheel bearings.
16. RJ The Bike Guy—26 April 2012—Second in Google search https://www.youtube.com/watch?v=IrzlHjhckSU. RJ created a second video undertaking much the same task which was uploaded to YouTube on 7 March 2014; it comes top of the Google search (https://www.youtube.com/watch?v=_axwV6sfaAs). The video quality is better and the work is closer to the camera and the voice closer to the microphone. It suggests that RJ has improved his technique in making repair videos—but the newer video does not serve my purposes of contrasting video styles here.
17. Third in Google search https://www.youtube.com/watch?v=LNzN9oTQA8s.
18. Fifth in Google search https://www.youtube.com/watch?v=yHvpzlcKBKg. The video carries the logo of What Mountain Bike, an international print magazine also published online through Magzter.
19. ATG is a UK organisation that provides apprenticeships and training schemes including 'cytech' courses http://www.atg-training.co.uk/cycle-training.html.
20. Sixth in Google search https://www.youtube.com/watch?v=wWEoGm1XY7U.

References

Black, John, Nigar Hashimzade, and Gareth Myles. 2012. *A Dictionary of Economics*, 4th ed. Oxford: Oxford University Press.

Dant, Tim. 2005. *Materiality and Society*. Maidenhead: Open University Press.

———. 2012. *Television and the Moral Imaginary: Society Through the Small Screen*. London: Palgrave.

Ellis, John. 2002. *Seeing Things: Television in the Age of Uncertainty*. London: I. B. Taurus.

Gelber, S. 1997. Do-it-yourself: Constructing, repairing and maintaining domestic masculinity. *American Quarterly* 49 (1): 66–112. Retrieved from http://www.jstor.org/stable/30041567.

Graham, Stephen, and Nigel Thift. 2007. Out of order: Understanding repair and maintenance. *Theory, Culture and Society* 24 (3): 1–25.

Gregson, Nicky, Alan Metcalfe, and Louise Crewe. 2009. Practices of object maintenance and repair: How consumers attend to consumer objects within the home. *Journal of Consumer Culture* 9 (2): 248–272.

Heidegger, Martin. 1962. *Being and Time*. Oxford: Blackwell.

Henke, Christopher R. 2000. The mechanics of workplace order: Toward a sociology of repair. *Berkeley Journal of Sociology* 44: 55–81.

Jonas, Hans. 1984. *The Imperative of Responsibility: In Search of an Ethics for the Technological Age*. Chicago: University of Chicago Press.

Leadbeater, Charles, and Paul Miler. 2004. *The Pro-am Revolution: How Enthusiasts are Changing our Economy and Society*. London: Demos.

Sartre, Jean-Paul. 2004. *The Imaginary: A Phenomenological Psychology of the Imagination*. London: Routledge.

Watson, M., and E. Shove. 2008. Product, competence, project and practice DIY and the dynamics of craft consumption. *Journal of Consumer Culture* 8 (1): 69–89. https://doi.org/10.1177/1469540507085726.

11

Repair for the Masses? Gender and Care Work in the Fixers' Collective

Meg Young and Daniela K. Rosner

Introduction

Victoria[1] worked hard to get her questions answered when she entered the back room of a local True Value Hardware Store, the monthly meeting site for Seattle's Phinney Ridge Fixers' Collective. Since 2011, Fixers' Collectives had offered Seattle residents public, cost-free events to meet with experienced fixers to repair broken consumer products, from vacuums and coffee makers to iPhones and laptops. For Victoria, a frank yet warm older woman, this site also served as a resource for her own project. She had brought a cracked cast-iron lamp. An ornate part at its top had come apart, but this was only a small break; not enough to

M. Young (✉)
Information School, University of Washington, Seattle, WA, USA
e-mail: megyoung@uw.edu

D. K. Rosner
Human Centered Design and Engineering, University of Washington, Seattle, WA, USA
e-mail: dkrosner@uw.edu

I. Strebel et al. (eds.), *Repair Work Ethnographies*,
https://doi.org/10.1007/978-981-13-2110-8_11

snap it off. As she worked, a retired Boeing engineer 'fixer' sidled up to her, explaining at length how she should epoxy it back together—how to hold it as it dried, and so on. But as he had been explaining the steps involved, Victoria interjected sharply: "*Gluing* is something I understand. It's *this* part I need help with," gesturing to the socket. The fixer accompanied Victoria to the store aisle where they might find the relevant socket switch, later reporting that the store did not have the right size. The fixer recommended that Victoria go to Harrell's, a specialty lamp store, to look for a new socket and the part housing it. As she left, she said she did not want any help carrying the heavy lamp to the car (May 6, 2015).

Around the same period, a Repair Fair convened some 200 miles south of Seattle in the parking lot of the local library in Cornelius, Oregon. Sophie, a shrewd, redheaded mom adept at tinkering with her own electronic appliances, brought her family from nearby Portland to repair a broken hair straightening iron. As Sophie explained her diagnosis of the problem to a male fixer volunteer, she said that she had a "*very* basic" understanding of how to repair it. She went into some depth about how she had already disassembled and examined it, describing how the wires had not been burned or melted. In a quiet, matter-of-fact way, she seemed proud of her diagnostic work. She had been reading a physics textbook to better understand the workings of things around the house she has tried to fix. A few weeks earlier, she had been able to fix an electric kettle by taking the bottom off to repair a circuit. The volunteer fixer helping her, a slight, neatly coiffed man in his late 70s, struggled to diagnose the problem. As he grew increasingly frustrated, Sophie tapered her claims of repair competency, as though not to undermine the expertise of the volunteer assisting her. She consistently punctuated her remarks with caveats like "I don't know enough about electronics" and "I tried to, I mean, but I didn't have much luck" (June 13, 2015).

In these settings, Victoria and Sophie variously assert and disclaim a technical identity. Though Victoria seeks neither chivalry nor gluing advice, she still wants help with the socket repair. Moments after declaring a fix easy, Sophie minimizes her ability. Both women asserted technical competencies, yet readily drew limits around these abilities. From standing warily in the entrance of the Collective, to saying that they do

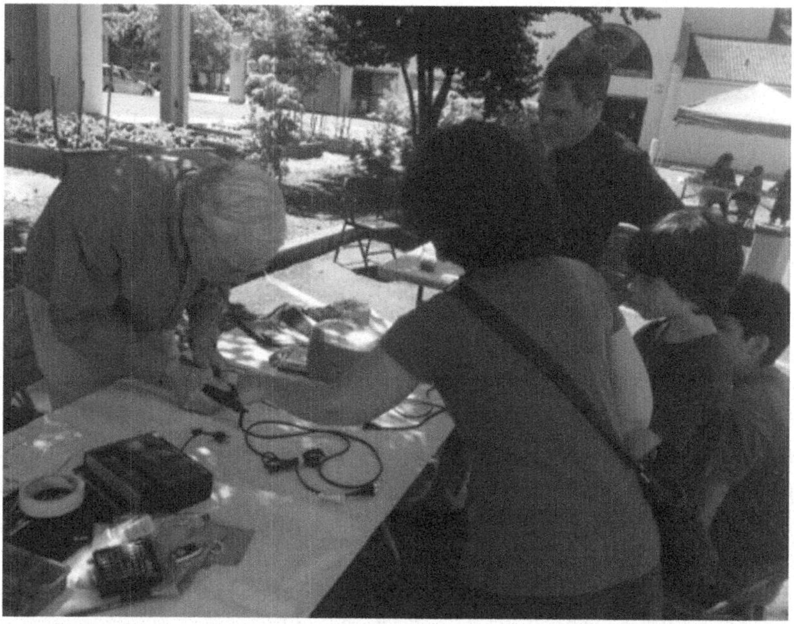

Fig. 11.1 Sophie explaining the disassembly of the straightening iron to the fixer volunteer (© Photo: M. Young)

not know where to start in disassembling a technology, female attendees paused before fixing. In our informal conversations and in our observations, women attendees often hesitated to open up their broken items. After months of fieldwork, accounts of visible intimidation littered our field notes (Fig. 11.1).

But why? Given the range of technical, enthusiastic, and engaged women involved in the parent organization of the Fixers' Collective, how do so few participate in repair? As part of a growing amateur repair movement in the United States and Europe, the Fixers' Collectives sought to connect fixing with a wide range of personal and collective programs, from environmentalism to hands-on tinkering. Organizer Kyle Wiens of iFixit, an online resource for electronics repair, has described using female hands in instructional videos to encourage participation from women (Wiens 2013). This parallels Amy Bix's (2009: 41, 44) description of photos from 1947 that pictured manicured fingers installing faucets before a post-war revival of gender demarcation. However, as Dunbar-Hester (2014: 54) has recently observed of ham radio activists, the relations between gender production and technical identity are "thorny and not easily overcome by good intentions." The fixers' efforts to unseat traditional gender roles come with moments of reinforcing those very same divisions.

To consider this challenge to demarcated gendered roles within the Fixers' Collective, we look to the growing body of scholarship on care and sociotechnical change. Spanning ethical philosophy, feminist epistemology, sociology, and science and technology studies, care ethics has traditionally focused on relationships between persons. Tracing care ethics across fields, scholars have considered care relationships as a unit of analysis that has implications for moral philosophy, epistemology, and ontology. Since Nel Noddings' advancement of a care ethics tied to women, and Carol Gilligan's claim that young women reason from within an 'ethic of care' instead of an 'ethic of justice,' scholars have reframed care as the logic from within which one undertakes moral reasoning (Tronto 1987). Tronto finds that Gilligan's argument is more subtle; not gender-determined, but gender-associated. Gilligan questioned the naturalization of gendered work, stating, "women's different moral expression might be a function of their subordinate or tentative social position" (Tronto 1987: 649). Thus care ethics scholarship offers us tools for interrogating received notions about gender and gendered competencies.

But what happens when we turn this lens on the repair of things? The range of toasters, coffee makers and pot lids coming into the Fixers' Collective foreground new material relations when viewed as acts of care. Whereas Gilligan and Noddings focus on care as a set of moral commitments, work within science and technology studies foregrounds the epistemological and ontological commitments care affords. In his recent call for 'broken world thinking,' Steven Jackson (2014) draws on care ethics to describe the mutual entanglement of people and things by which moments of collapse, disintegration, maintenance, and repair unfold. "To care for something is to affirm a moral relation to it" (Jackson 2014: 231–232). Drawing on Haraway's feminist readings of situated knowledge, de la Bellacasa (2012) takes this moral relation further to cast care

Fig. 11.2 Northeast Seattle Tool Library Fixers' Collective (© Photo: M. Young)

as a way of understanding objects laden with interdependency wherein, in Haraway's words, "Nothing comes without its world," (de la Bellacasa 2012; Haraway and Goodeve 2018: 137) (Fig. 11.2).

Within the Fixers' Collective, framing care as a livelihood skill complicates the durability of feminized, invisible or unpaid maintenance work. As radically localized events, processes of care and maintenance take shape as heterogeneous relations between gender and technical competence. An attention to care has the power to make visible those multiple and overlapping competencies often overlooked. In doing so, it may highlight labor conditions and approaches connected with domination (de la Bellacasa 2012). But rather than use care as a way of seeing the devaluing of feminized work, we use care to highlight "an ontological requirement of relational worlds" (ibid.: 199). That is, care becomes fundamental to how programs of gendered work come to be, and get maintained in relation to one another—in the Fixers' Collective and beyond.

These concerns reverberate in a growing body of ethnographic work on repair. If we were to look back at the last three decades of repair studies, the kinds of groups that participate in repair may appear to differ widely along particular dimensions and look more similar along others. Technicians tinker with new broken audio sensors in Kampala (Houston in this volume, Houston and Jackson 2016); auto mechanics grease the chain of old motorbikes in the US and UK (Dant in this volume); and maintenance workers replace signs in the Paris subway (Denis and Pontille in this volume). These accounts touch a wide range of geographies and programs. Yet, they also primarily foreground the work of men.

As we look beyond repair to an emerging literature on amateur and Do-It-Yourself (DIY) technology development, new possibilities for rewriting the gendered scripts of technology development emerge. Recent accounts of DIY home improvement (Moisio et al. 2013), ham radio activism (Dunbar-Hester 2014), feminist hacking (Fox et al. 2015) and American "chicks who fix" (Bix 2009) offer analytical leverage for complicating masculinized histories of technical labor. They detail how the gendered demarcation of technology development has developed with and through an idea of home improvement as a masculine pastime. While Bix (2009: 40) describes the expansion of American professional home economics in the 1920s as establishing "its own gendered technical expertise in forms defined by and for women," by the end of World War II domestic repair became, in her words, "remasculinized." As Steven Gelber asserts, "the very term 'do-it-yourself' would become part of the definition of suburban husbanding" (Gelber 1997: 67, also quoted in Bix 2009: 38). In her recent accounts of ham radio activists, Dunbar-Hester (2014: 55) shows the entrenched nature of this gender disparity as a notion of masculine technical prowess, and the ways it has endured activists' attempts to unseat it. Repair, in these worlds, emerges from the shadows of a masculinized labor politics.

Here we build on Dunbar-Hester's (2014) accounts of radio activists and Bix's (2009, 2014) histories of women fixers to suggest that emerging sites of amateur repair offer alternative configurations of gender identity. Drawing on fieldwork with amateur fixers in the Pacific Northwest (Houston et al. 2016) and Northern California

(Rosner 2014; Rosner and Turner 2015), we consider how studies of technology repair might grapple with a shifting gender politics. We focus on Seattle's Fixers' Collectives, events that began with the rise of tool libraries in the 1970s. Like the role conventional libraries have played for books and communication media, tool libraries enable local residents to temporarily borrow household appliances and implements such as lawnmowers, shovels, and drills. By becoming home to Fixers' Collective events, they framed fixing as a communal good to develop public theaters of technology production (Rosner and Turner 2015). But they also complicated conventional relationships established around technical expertise and gender.

This chapter turns to Seattle's Fixers' Collectives at three sites—the Northeast Tool Library, the Phinney Ridge Tool Library, and the West Seattle Tool Library—to explore what these alternative modes of production offer repair studies. In particular, we use this fieldwork to address two kinds of questions. Empirically, we consider the kinds of gender dynamics these sites help make possible. For example, do they help to challenge established gender roles, or do they help reinforce them? Analytically, we ask how our engagements with care and repair can help us produce different accounts of gender, and how our engagements with gender can, in turn, help us develop different ways of understanding care and repair. In recognizing these engagements, this chapter charts opportunities for examining the forms of fixing that reinforce, build or undo individual and collective gender identity.

The Coming of the Fixers' Collective

Our study is based on fieldwork conducted in hobbyist communities of repair work in Seattle, WA and surrounding communities. Since the mid-seventies, repair hobby work in Seattle has occupied community-organized public sites called tool libraries. Tool Libraries allow members to check out hardware, gardening, and other home tools much like communication media and books in conventional libraries. Whereas some tool libraries have closed their doors, others, such as that within the Phinney Neighborhood Association, have operated continuously

Fig. 11.3 Workshop space of the West Seattle Tool Library (© Photo: M. Young)

since the late 1970s. After the establishment of Phinney, two neighborhood organizations concerned with "green" living, called Sustainable West Seattle and Sustainable Northeast Seattle, opened their own tool libraries in 2009 and 2012 respectively. These sites have sustained membership, donations, and circulation over the years, often hosting repair groups called 'Fixers' Collectives' (Fig. 11.3).

These repair groups drew their name from across the country Brooklyn, New York, where in 2008, a group of artists started a social experiment called the Fixers' Collective in a gallery. The idea was so successful that it took shape as a regular event. A few years later, a regular volunteer at the West Seattle Tool Library was reading about the Brooklyn-based group online. Gary Sato, a craftsman, had recently bought a salad spinner whose mechanism had broken within the first few weeks of use, leaving him feeling the weight of disposability in "*things that are just not really meant to be repaired,*" as he put it. While he had no formal ties to the Brooklyn Fixers, Gary began a Fixers' Collective in the spacious, public woodworking shop of the West Seattle Tool Library in the summer of 2011. The West Seattle Fixers' Collective was very active in its initial years. In early 2013, one of the organizers, Daniel Brown, moved to the Phinney Neighborhood Association

Tool Library and started a new Fixers' Collective there, as a way to promote the tool library to the wider community. Soon after, the Northeast Seattle Tool Library, too, started a Fixers' Collective as a way to meet their sustainability mission and increase membership:

> At PNA they just have space for the tools. You walk in, you get the tools, and you leave. The [West Seattle] Fixers' Collective was this great opportunity to hang out, basically, to get to know some of the users of the tool library, and also just welcome in other people, just create an unofficial welcome, rather than saying that you have to come on in and check things out and use the service… That's how it got started [at Phinney]. It's just a way to create community for the tool library. (Brown, June 17, 2015)

Each of Seattle's Fixers' Collectives connect to their respective tool libraries and neighborhood organizations in terms of recruiting attendance and regular volunteers.

Across our three Seattle field sites, the format of Fixers' Collective events looks similar. A cadre of fixer volunteers meets one night a month, prepared to help attendees with their repairs. Two of the Collectives meet in the workshop space in their own tool libraries, a third meets in the back of a local hardware store. Although the time and date of the meeting are formalized, Seattle's repair events are informal. They are announced online, but not widely promoted. In each Collective, one volunteer has taken responsibility to attend each week. Despite meeting over 25 volunteers across our half year of fieldwork, every volunteer we observed in Seattle was male, most often Caucasian, and in retirement from an engineering position at a local tech or aerospace engineering company. Daniel, who had been a driving force in organizing two of our field sites, attributed this dynamic to the hardware store culture of the tool library as the original setting in which the Collectives were organized.[2] Seattle's fixer volunteers were known as 'fixers,' repair enthusiasts who come to Fixers' Collectives to assist attendees. While fixers were primarily men, our other sites in Portland and the Bay Area pointed to other possibilities for who 'fixers' are or can be. Here, we refer to those who attend Collectives regularly as 'fixers,' and those who bring items in need of repair as 'attendees.'

Accounting for the Participation of Women and the Active Fixer

Early in our fieldwork we found organizers of events reflexively aware of gendered divisions of labor. We learned that the West Seattle Tool Library used the Fixers' Collective to host a monthly women-focused series on using power tools and home repairs. Alice Jones, a carpentry student in her mid-20s who has tattoos of bike wrenches and a wood cabin, is the new coordinator of the Northeast Seattle Tool Library. She hoped to teach a similar series in a forthcoming season (Fig. 11.4):

> I really just want there to be an easy informal class so that everyone can do this and doesn't have to feel like 'you're a woman, so here's your like flower [patterned tools]'—you know those hammer sets that have flowers on them? I want to have an informal, just like "here's how to use a circular saw, it's not actually that scary once you learn how to use it." It's better than having someone say, [adopts a deep voice] "What are you scared about? Just turn it on!!" It's like, no, it will cut off your fingers if you use it wrong; that's a real thing, it happens today. It could happen to anybody. There is a mansplaining mentality that goes around tools. It's kind of the reason I did the carpentry program, cause I was tired of either feeling stupid or feeling like treated different 'cause I was a woman trying to use tools. (Jones 2015)

Many of the women who participated in these sites were already highly technical, or eager to learn more. Connected to this framing is the way fixers framed their work as empowerment: giving new control to consumers by instilling an interest in developing repair competencies. To explain this, fixers often recalled their 'first repair' stories of successful fixes and described forms of self-efficacy that developed from these achievements. However, this narrative of empowerment came with selective forms of agency and aspirational metaphors that complicate the story of who controls repair work in practice.

When asked about his favorite fixes to date, for example, soft-spoken academic and fixer volunteer Jonathan Escher recalled a mother and daughter who brought in Kitchen Aid standing mixer. The motor did

Fig. 11.4 Tape measure branded as 'Pretty Tools, Tools Designed with a Woman in Mind' (© Photo: M. Young)

not run when it was switched on. "They just felt so powerful because they had restored this thing to new condition," he said. "We hand them tools, just like we would do with the men, and urge them, 'Don't worry, it's already broken.' And they just dive in" (Escher 2015). The month before our interview, Jonathan had worked with a shy woman in her late 40s, Lillian, who entered the PNA Fixers' Collective, wearing a hoodie and ball cap. Lillian had an old clock radio; the metal prongs at the end of the electrical cord had snapped off. The Collective was particularly busy at that moment; she stood cautiously at the boundary of the space, waiting her turn to enter and reticent to claim space at the work table. She brought in the clock radio for repair, but she seemed to be intimidated and unsure how to proceed. It appeared simple to fix, and I (Young) told her that we should be able to set her up with a wire stripper to get started. She asked me, "Are you an engineer?" Put off by this comment, I introduced her to Jonathan, who took the clock radio in his hands and began to pull apart the plug himself, asking me for a wrench. Although he talked Lillian through the repair as he went, he proceeded to strip the wire and replace the end of the cord himself. Lillian stayed until the end of the Fixers' Collective, watching the fixers disassemble two old machines they had brought into tinker with (a circuit tester from the 1940s, and an electronic Simon game from the 1980s). Even as the crowd thinned, she stood physically apart from the rest of the fixers, leaning into observe from a few feet away (6 May 2015). Here we see how fixers often took up the work themselves, despite efforts to give attendees control over their repairs (Fig. 11.5).

Sometimes, attendees demarcate their difference from the fixer volunteers via a pronounced gratitude for their assistance. In the PNA Fixer Collective, a vivacious woman in her mid-50s brought in a broken kitchen chair. She dealt graciously with the fixers compared with other attendees, often thanking them profusely. "Getting all this attention, I feel so special now!" she exclaimed (May 6, 2015). In NESTL, another woman intermittently interjected her thanks to the volunteers throughout the repair process. When a fixer took her broken object and finished it, she looked around the room and let out a loud gasp to demonstrate how pleased she was with the fix. At an event we observed in Northern California, a young entrepreneur had attended many fixers events, but still marveled outloud at each step that the fixer helping her made with her humidifier. She combined this work with loud asides about how she

Fig. 11.5 As she lingered, she stood physically apart from the rest of the fixers, leaning into observe from a few feet away (© Photo: M. Young)

would have "*No idea* where to start" if she was doing the fix herself, or what to do at each step. One would not know from this performative stance that this attendee was herself a prominent local organizer of repair-related community events. We interpret these gestures as a form of emotional labor, by which attendees supported those around them (Hochschild 2003).

This pattern continued in the Northeast Seattle Fixers' Collective. Kay, a slender woman in her early 60s with long gray-blonde hair, brought in a lamp with a broken switch. She was an active member of two of the Fixers' Collective umbrella organizations, Sustainable NE Seattle and its tool library; she visited the Fixers' Collective regularly. While some attendees viewed themselves as a 'customer,' Kay described her membership in the community created by the Collective. The day we met, she asked for a soldering iron to fix a lamp she had brought in, explaining that she did not have one at home. At this point, Amin, a wiry, animated retired engineer in his late 60s swept into the conversation, single-minded about his mission. Saying little, he located and started setting up the soldering kit. Kay remained at his side, putting on gloves and holding the part in place, while the fixer soldered the lamp back together. Whereas Kay had signaled that she was planning to take up the project alone, the fixer took the act of repair upon himself.

Alongside these setbacks for female attendees, male attendees also benefited from intimate instruction. Richard, a mild man in his late 70s, brought in a power saw to be repaired. He was assisted by Bo Presley, a robust union carpenter and fixer volunteer in his early 60s. Unlike Kay, Richard meekly identified himself as a 'customer' and seemed to expect Bo to do the fix for him. Rather than take control, Bo took pains to walk Richard through the saw's disassembly even when Richard did not seem interested. Bo identified the switch as the problem, instructing Richard as to where to find a replacement, and how to install the new switch after buying one. Later, talking to the researcher, Bo used this as an example of an ideal fix. "See, with Richard, I was not doing it behind his back—I was showing him, in front of him like this." He illustrated differences in body language—carrying out the repair with a closed stance versus an open and demonstrative one. "We don't just fix it and hand it back, we make them do it. It's the old fish thing.

We believe in teaching people to fish" (May 5, 2015). Bo's comparison to fishing evokes a sense of repair as a primal livelihood skill. In doing so, he reinforced an idea of technical masculinity that can be traced back through American Progressive Era home improvement, in which developing technical mastery comprised an important part of male identity-formation (Douglas 1999; Dunbar-Hester 2014; Bix 2009).

Reflecting this same imagery, fixers often cited the archetype of one repairman-rebel in particular: Harry Tuttle of the dystopian film, Brazil (1985). As a rogue heating engineer who repairs units illegally, thus defying the state, Tuttle gave male participants in multiple field sites a useful shorthand for "a real revolutionary," as one fixer explained (May 5, 2015). "It's just the idea of, a world where things are broken," Jonathan explained of the film. "It's just the image of repair being this *illegal activity*, this outlaw activity so that you've got this outlaw hero" (Escher, June 19, 2015). For male fixers, repair was not just an act of transgression, but also a partial reclamation of ownership of technologies that (at times) seek to thwart it in the proprietary designs of parts like screws and bolts.

Here the act of fixing corresponds to mechanisms of seizing control. Repair gained a heroic valence from the fixers' sense of fighting off larger forces, which constrain or inhibit individual agency. Despite aiming to play a supporting role to women's agency-formation, fixers had trouble rejecting masculinized ideals of mastery over technology. For those who provided and performed their expertise for attendees, the durability of these gendered scripts grew stronger. However, as we see in the relationships of care built into objects, an attention to care may make present the gendered competencies and labor conditions often hidden in acts of repair. As we see in the stories that follow, care relations become important to explaining the many processes at work in the collective, including the production and maintenance of gender distinctions.

"Good as New" Repair

In the setting of the Fixers' Collective, a large intangible goal like 'empowerment' gained tangible expression in how participants diagnosed the fix. On first blush, fixers appeared to welcome attendees to bring in any kind of object, from CRT monitors to Kitchen Aid Mixers, or chairs and steel pans. To better understand the range of problems fixers value, consider the repair diagnosis in Phinney Ridge, when a senior Filipino man brought in a laptop, explaining that he had forgotten his password and been locked out of his computer. Although none of the fixers knew how to unlock the computer, they nevertheless took the laptop and turned it on, spending time restarting it a few cycles before advising the couple on where else they could find help. Displaying similar commitment, a man who brought in a rusted shovel to the NE Seattle Collective met a fixer volunteer who gave the attendee advice on how to sharpen the blade and sand down the handle.

Upon closer inspection, however, subtle signals revealed decisions by which fixers welcome some repairs over others. During an event at the Northeast Seattle Tool Library, I (Young) brought in a piece of old costume jewelry. A fixer volunteer, Alex, a large, outspoken man in his 50s, picked up the pieces and jeered, 'You want to fix *this*?' A month later, the same fixer volunteer helped Kay, the regular Fixers' Collective attendee, to repair a broken pot lid that she brought into the Northeast Seattle Fixers' Collective. The pot once had a glass lid which had fallen and shattered—just its metal rim and knob were left. Kay had attended the Fixers' Collective many times before and was buoyed by her plan to repair the lid by cutting a disposable aluminum turkey baster to fit where the shattered glass had been. Alex assisted her, as described in this excerpt from our field notes:

> Alex was giving her kind of a hard time about the project, which became more pronounced later. Alex also had a kind of showmanship, going into a film noir voice in describing the repair. Together, Alex and Kay cut a circumference from the aluminum that fit the pot lid, and used rubber wedges and a mallet to fit the aluminum into the crevices of the lid without tearing it. There was an open question of how to get the aluminum to stay attached to the lid. Alex was adamantly advising her to use some kind of glue or caulk, saying it would be no big deal because "[she] wouldn't eat off of it anyway." Bo, nearby, was much more concerned, warning Kay that she needed to make sure it would be fire-rated silicone, if she used it at all, and saying he didn't think it was a good idea: "Don't

> put it in the oven!" "This one never goes in the oven," she said. Bo was also concerned that bringing different metals into contact with each other would cause accelerated corrosion. Kay had stepped out for a moment, and Alex was saying some things that were at first just veiled criticisms. He was laughing, mocking the project. "It will be a piece of art, knowing Kay. A piece of recycling." Then, something a bit more judgmental: "It's up to Kay if she wants to show her friends or not". (May 5, 2015)

In the excerpt above we see how Alex volunteered to aide Kay, then raised questions about the craftsmanship of the work when she stepped out of the room. He characterized the work as "art," and implied that her handiwork might be inelegant or embarrassing—something not to show to others. For Alex, a repair should restore an item to its original form. Otherwise, it is of diminished quality; as he added later, "I got to hand it to Kay. The pot will probably hold up through two uses." This parallels what Fox and Heinemann (2015) in their study of customers in a shoe repair shop describe as a tension between "knowledgeability" (what the customer understands of repair) and "grantability" (the service's ability to meet a customer's request). "[W]e couldn't guarantee what it'll feel like, but, if that's what you want we'll do it. Yeah," a shoe repair technician explained with apparent reluctance (ibid.: 355, removed analysis punctuation for readability). Alongside concerns for inclusion, fixers were keen to restore function to objects, and often crowded around electronic and mechanical repairs. Technical concerns sometimes downplayed or undermined other motivations for repairing an object, like sentimental value (Fig. 11.6).

Latent in this way of valuing objects is a sense that the finished product should look 'good as new.' Fixers take steps to minimize visible traces of repair and, in so doing, they could continue to shape women's participation in the space. When Chelsea, an effusive woman in her 50s, brought in a dining room chair to the Phinney Fixers' Collective, one man sprang to her side, saying, "Finally, something wooden!" He asked her how visible the chair was in her home, to guide his choices on how to fix its broken leg. "I'm not fussy!" she said, explaining that the chair was an inexpensive model from a grocery store. The fixer was distressed that the repair would result in a visible silver screw on its side. When

Fig. 11.6 Kay consults Guy (left) and Alex (right) (© Photo: M. Young)

Chelsea repeatedly assured him that the visible repair would not bother her, he remained unconvinced. He agonized over the visible screw marring the honey-colored wood. At times, his focus on traceless, 'good as new' repair began to undermine Chelsea, who had to minimize herself and the chair for him to proceed (May 6, 2015). Like Kay and her pot, Chelsea's way of relating to her chair was not contingent on it looking 'good as new.'

We observe a tension between these conflicting repair esthetics. In one formulation, repair restores an item to its original manufactured form. For fixers, traceless repair signaled technical competencies and mastery. Nevertheless, many fixes could not be completed without the telltale signs that a repair had taken place. In a different configuration of values, this visibility reveals beauty, because traces reflect acts of care that ennoble the objects that have been thus cared for. The Repair Manifesto of Sugru, a new adhesive marketed at craft and maker festivals, draws on this latter way of thinking about repair: "A fixed thing is a beautiful thing. Every fix, whether skillful or improvised, holds a story." This framing resonates with the words of Daniel Brown, a boat repair worker and the founder of Phinney's Fixers' Collective: "To be able to see something repaired and add that sentiment to it, actually *there's* the craft. Now I've worked on this plastic, chintzy, designed-to-break thing. Now it has that soul. I've just given it that craft, that soul, part of me. Now I own it. I think that's just a real beautiful thing. I think I can romanticize it quite a lot" (Brown 2015).

The form a repair takes, in this sense, could make visible these different ways of relating to the object. Approaching these relations as care work recovers their context, subtly complicating the gendered scripts of repair. Gender production and technical competence emerge as connected, heterogeneous undertakings, defined and reinforced through object-relations. These object relations, however, do not exist in a vacuum, but get built (and rebuilt) through care work. Care becomes core to how the asymmetry between attendees and fixers, or gender demarcations, come to assert themselves over time. In this way, care is not just a lens on repair, but constitutive of gender workings in the fixers' worlds.

Rethinking Gender Through Repair

So far we have considered the gendered demarcation of repair against the backdrop of Seattle's Fixers' Collectives. Three conceptual framings organized this fieldwork: the ways these groups dealt with questions of women's participation, the narrative of the active fixer, and the sometimes-patronizing discourse of "good as new" repair. These frames were deeply implicated in the formation of one another. The bounding of repair that surfaced in Alex's dismissal of Kay's pot lid, for example, reemerged in Lillian's question of Young, "Are you an engineer?" More broadly, we saw what Dunbar-Hester (2014: 66) described as "dissent from 'mainstream' values" among her ham radio activists: countercultural ideals that gave no assurance of denying traditional gender roles.

The break between the discourse and practice of including women became a key site of this difficulty. While volunteers often asserted women "don't watch," they instead "just dive in" (Escher, June 19, 2015), we observed only a handful of women attend Fixers' Collectives and even fewer "dive in." Depending on who was involved, the groups often naturalized repair as a universal practice. Yet this universalizing discourse provided a useful foil, enacting subtle reinforcements of repair as a core competency of men, and not women.

However, the Fixers' Collective itself played an interesting role in this gender production. By containing gender work and its expression within the confines of their events, fixers could make room for participants to challenge and undo strict divisions of labor beyond the Collective. Women attendees evangelized repair work during informal conversations at work and took up challenging repair projects on their own time. Sophie's proficiency to fix her hot water boiler or Daniel Brown's recognition of beauty and soul in repair suggests a reflective complicating of gendered scripts.

Beyond the construct of gender, other factors contributed to dynamics of inclusivity. For many, the Fixers' Collective represents new territory for community building. Be it a particular hardware store or a non-profit 'tool library,' the sites hosting Collective events introduced local residents to new faces and commercial interests. In visiting these sites, attendees navigated not only their unfamiliar values but also a need to ask volunteers for help, from signaling for attention to introducing themselves and presenting their problem. Moreover, most fixer volunteers came to Collective events already familiar with one another, while attendees often met everyone for the first time. Many unfamiliar processes were at work in reproducing gender as a salient factor—both in terms of who visited the events, and how visiting women took part in a repair.

By attending to the work of reassembling broken objects as acts of care, Fixers' Collectives defy conventional ways of understanding technological competencies. They help us step outside instrumental concerns for technology and gender to highlight the engineering decisions of industry ("chintzy" materials, "designed-to-break" circuitry), particularly as they meet up with local ecologies of attendees, fixers, and objects. Embracing an ontology of care, we depart from the reassertion of gender, technical and attendee/fixer distinctions to instead understand these relationships as nested, partial and performative. Mundane maintenance—made less ordinary in the context of the Collectives' monthly community workshops—entails not only encounters with tools, but also the continued cultivation of individual and collective care relations. This ethic invites an alternative reading of gender disparity that casts technical accomplishments as potential (often underutilized) mechanisms for undoing the gendered scripts of repair.

Notes

1. All names in this work have been pseudonymized.
2. Daniel noted that in the time before our fieldwork started, the Fixers' Collective in the West Seattle Tool Library "often ended up being 50/50 with gender and all different ages, also, people just coming to it for different reasons… Something I'd love to see keep developing at PNA would be a lot more diversity" (Brown 2015).

References

Bix, Amy Sue. 2009. "Creating chicks who fix": Women, tool knowledge, and home repair, 1920–2007. *WSQ: Women's Studies Quarterly* 37 (1): 38–60.

———. 2014. *Girls Coming to Tech!: A History of American Engineering Education for Women*. Cambridge: MIT Press.

Brown, Daniel. 2015. Interview with Meg Young, June 17.

de la Bellacasa, Maria Puig. 2012. 'Nothing comes without its world': Thinking with care. *The Sociological Review* 60 (2): 197–216.

Douglas, Susan J. 1999. *Listening in: Radio and the American Imagination*. U of Minnesota Press.

Dunbar-Hester, Christina. 2014. *Low Power to the People: Pirates, Protest, and Politics in FM Radio Activism*. Cambridge: MIT Press.

Escher, Jonathan. 2015. Interview with Meg Young, June 19.

Fox, Barbara A., and Trine Heinemann. 2015. The alignment of manual and verbal displays in requests for the repair of an object. *Research on Language and Social Interaction* 48 (3): 342–362.

Fox, Sarah, Rachel Rose Ulgado, and Daniela Rosner. 2015. Hacking culture, not devices: Access and recognition in feminist hackerspaces. In *Proceedings of the 18th ACM Conference on Computer Supported Cooperative Work & Social Computing*, ed. ACM, 56–68.

Gelber, Steven M. 1997. Do-it-yourself: Constructing, repairing and maintaining domestic masculinity. *American Quarterly* 49 (1): 66–112.

Gilliam, Terry, Charles McKeown, and Tom Stoppard. 1985. "Brazil." Feature film, Embassy International Pictures NV, distributed by Universal Pictures, US.

Jackson, Steven J. 2014. Rethinking repair. In *Media Technologies: Essays on Communication, Materiality, and Society*, ed. Tarleton Gillespie, Pablo J. Boczkowski, and Kirsten A. Foot, 221–239. Cambridge: MIT Press.

Jones, Alice. 2015. Interview with Meg Young, August 4.

Haraway, Donna J., and Thyrza Goodeve. 2018. *Modest_Witness@ Second_Millennium. FemaleMan_Meets_OncoMouse: Feminism and Technoscience.* Routledge.

Hochschild, Arlie Russell. 2003. *The Managed Heart: Commercialization of Human Feeling, with a New Afterword.* Berkley: University of California Press.

Houston, Lara, and Steven J. Jackson. 2016. "Caring for the next billion mobile handsets: Opening proprietary closures through the work of repair." In *Proceedings of the Eighth International Conference on Information and Communication Technologies and Development*, ed. ACM, 10.

Houston, Lara, Steven J. Jackson, Daniela K. Rosner, Syed Ishtiaque Ahmed, Meg Young, and Laewoo Kang. 2016. "Values in repair." In *Proceedings of the 2016 CHI Conference on Human Factors in Computing Systems*, ed. ACM, 1403–1414.

Moisio, Risto, Eric J. Arnould, and James W. Gentry. 2013. Productive consumption in the class-mediated construction of domestic masculinity: Do-It-Yourself (DIY) home improvement in men's identity work. *Journal of Consumer Research* 40 (2): 298–316.

Rosner, Daniela K. 2014. Making citizens, reassembling devices: On gender and the development of contemporary public sites of repair in Northern California. *Public Culture* 26 (1): 51–77.

Rosner, Daniela K., and Fred Turner. 2015. Theaters of alternative industry: Hobbyist repair collectives and the legacy of the 1960s American counterculture. In *Design Thinking Research*, ed. Hasso Plattner, Christoph Meinel, and Larry Leifer, 59–69. Berlin: Springer.

Tronto, Joan C. 1987. Beyond gender difference to a theory of care. *Signs: Journal of Women in Culture and Society* 12 (4): 644–663.
Wiens, Kyle. 2013. Conversation with Daniela Rosner, April 28.
Young, Margaret, and Daniela K. Rosner. Under review: Procedural vision: The decline of tech activism in Free Geek Seattle. *ACM Conference on Human-Factors in Computing*.

Part IV

Afterword

12

Repair as Transition: Time, Materiality, and Hope

Steven J. Jackson

"Life is in the transitions, as much as in the terms connected."

—William James, 1904

The essays that precede this one represent the first serious scholarly effort to put repair practice into comparative perspective. Richly empirical and drawn into conversation through an extended collaborative process, they offer both insight and raw material for reevaluating the nature and status of maintenance and repair work across a wide range of sites and contexts today. They may also help us to think differently around questions near and dear to the heart of the modern social sciences, providing new inroads into key issues that challenge both scholarship and action in the world. Wide-ranging, empirically grounded and undisciplined (in the good sense): what lessons can we take from such studies as a set?

S. J. Jackson (✉)
Cornell University, Ithaca, NY, USA
e-mail: sjj54@cornell.edu

I. Strebel et al. (eds.), *Repair Work Ethnographies*,
https://doi.org/10.1007/978-981-13-2110-8_12

The first and most obvious concerns the remarkable range of sites and conditions under which repair work operates. From book restoration in an Austrian library to signage in the Paris metro, the repair of mobile phones in Kampala to the recuperation of broken medical devices and practices in the US, the essays here remind us that repair figures as an ever-present feature of the sociomaterial world around us, including (perhaps especially!) in places where we've tended to neglect or misrecognize its presence. A similar range characterizes the scale at which repair can be seen to operate, from the immediacy of "settings" addressed in part one, through the "networks" and "politics" central to the arguments of parts two and three. In all cases, as the essays remind us, the durability of social and technical orders is neither given nor guaranteed through moments of design or construction. Things—even clever and well-assembled ones—rarely stay put. Human practices are similarly unruly, prone to all sorts of failures and departures that demand constant attention and recuperation: as Melville (1851) reminds us, "of all tools used in the shadow of the moon, men are the most apt to get out of order." The essays here show us this work in action, calling to light the careful and ever-present work by which order and durability, wherever they are to be found, are constituted and sustained.

The essays also speak to the situated character of repair work and knowledge: its deep embedding in contexts that structure the need and possibility of repair, and furnish the resources available for its accomplishment. In this view, repair work is deeply contingent, dependent on social and material configurations grown and evolved in particular time and places. Much of the diagnostic work embedded in repair is in fact about calling these configurations to light, both to understand the nature of breakdowns and problems encountered, and to craft effective and doable responses from the resources that context affords. Several of the pieces also tackle the problem of how repair knowledge travels and is shared, whether through structured workplace environments (Schubert's medical professionals; Tironi's bicycle repair crews) or more diffusely through forms of online exchange and informal networks seen in Dant's bicycle repair tutorials, Houston's Kampalan mobile phone technicians, or the participants in Young and Rosner's amateur fixer's collectives.

A third point, more subtle and elusive, concerns the grounding of contemporary repair studies in longer theoretical and philosophical traditions—and conversely, the deployment of repair as a lever or spring (perhaps screwdriver?) with which to "fix" and extend received bodies of social scientific thought. Schubert's study of medical breakdown and repair for example shows the deep roots of repair sensibilities in pragmatist philosophy and interactionist sociology, suggesting how Deweyan notions of inquiry, routine, and improvisation might be inflected to provide new and generative descriptions of repair work in medical and other contexts. Denis and Pontille (re)connect repair to theories of assemblage growing at the three-way intersection of geography, urban studies, and actor-network theory, showing the insights that repair can bring to other connection-centered approaches. In both classic prior work (Henke 2000) and in his contribution to this volume, Henke argues for repair as an ever-present feature of social and infrastructural life, moving from static and structuralist conceits to a relational model which takes seriously both the fragility and power of inherited infrastructures, and the role of repair in sustaining or undermining them. The editors' introduction further elucidates these connections, illuminating the varying pragmatist, ethnomethodological, and Heideggerian traditions that might reasonably be claimed as ancestry by contemporary repair scholars.

Finally and most concretely, the essays speak to the distinct forms of skill and agency involved in repair work, and the remarkable suppleness with which such skill is frequently deployed. We see for example the innovative and collaborative processes through which workers and volunteers in Lejeune's steam engine trusts learn to diagnose and straighten rods, tighten cables, and replace blowers to maintain the viability of vintage steam engines (a process that as Lejeune correctly points out involves both a going back and going forward in time). We also see the various kinds of hacks, kluges and workarounds that support repair in amateur and enthusiast environments, along with creative modes of sharing (Dant) and intersections with gendered technical identities (Young and Rosner) that shape and sometimes complicate repair work.

In Denis and Pontille's study of signage in the Paris Metro, maintenance emerges as a kind of "dance" through which human and material flows are drawn into artful and workable assemblage, all within the workplace logics of the metro authority and the dynamics of urban form.

That any of this needs stating, of course, speaks to other features endemic to repair, as well as the particular (and particularistic) kinds of stories we have learned to tell about the worlds around us. As the editors rightly observe, much has been made of the invisibility of repair work, and its tendency to disappear from view in the context of smoothly functioning systems. From Heideggerian emphases on breakdown (and his famous example of the broken hammer [Harman 2002]) to more recent strains of work in infrastructure studies (Edwards et. al. 2007, 2009), a long line of scholars has attended to the distinct challenges of visibility that confront repair and other forms of infrastructural labor, and conversely, the particular forms of recognition and disclosure available when they break down. This insight has been mobilized methodologically (as for instance in Garfinkel's (1967) famous "breaching experiments"). It has also been deployed politically, as a reminder that the invisibilities assigned to maintenance and repair connect as well to histories of valuation that raise some forms of work (and workers) above others, and that are themselves linked to broader social concerns around class, race, gender, and several other powerful stratifications. Under such circumstances, repair studies like the ones in this volume offer both counter-history and double vision, giving us new insight into the objects and systems around us (and the work required to keep them live in the world): buildings as entities shaped and sustained by maintenance staff, not (just) architects; mobile phones as the products of fixers, not (just) designers. As the editors observe, this kind of analytic and political switch is central to the strategy of infrastructural inversion advocated by Bowker and Star (Bowker 1994; Bowker and Star 1999). It also points to different and neglected facets of our relationship to technology and the material world around us. While the authors adopt different ways of naming these relations—bricolage, inquiry, care, etc.—the broad impetus is shared: that there are different ways of being *in* and *with* the material world through time, and which careful consideration of repair helps call to the fore.

Present in the individual studies, these lessons come out all the more powerfully through the comparative possibilities that the edited collection allows (at least when supported by the sustained collaborative engagement and editorial guidance featured here). Comparative work at its heart lies in the effort to *think things together*, in their sameness and difference, building out the forms of connection, disjunction, and adjacency that together form the shared-but-discrepant spaces that are the real-world terrain of contemporary scholarship and practice. A part of this contribution operates at the level of comparative empirics and the forms of theorizing that may follow from them. But another part is wider and more diffuse in nature, speaking to the basic sensibilities and worldviews through which we apprehend and organize the world of experience (including what and who we choose to take note of, and what dynamics and phenomena we choose to find interesting). Taken as a set, it is here that we find the most profound contribution of the assembled pieces: beyond their empirical and theoretical contributions, the pieces here invite us to *imagine repair differently*—as practical, theoretical, and perhaps political category. This is not just about "seeing" new things, filling in blank spots on the received scholarly map (though these are plentiful, and the essays here cast light on several of them). It is also about *seeing old things differently*, and using repair as an invitation to rethink (indeed, fix!) received bodies of work, in STS and the social sciences more generally. Whether sensibility, worldview, or "broken world thinking" (Jackson 2014), the essays here suggest ways of understanding and participating in the flow of events and objects around us that other forms of naming—including many core to the metaphors and imagination of the inherited social sciences—tend to forget, misrecognize, or categorize away.

In all these ways, the essays in this volume help to gather and define the vibrant but still-nascent field of repair studies. They show us the range of repair work, its presence and manifestation across a wide array of sites and contexts. They show us something of its stakes: both in regard to the systems, institutions, and infrastructures that rely on it, and in the lives of groups and individuals who perform it. And they show us something of its theoretical roots and attachments, suggesting in the process problems and possibilities that no doubt merit (and will receive) further development going forward. In the remainder of this essay I speculate on three of these, and suggest how empirical and theoretical work around repair may offer fresh perspective on themes and problems that have moved increasingly to the center of STS and broader social science concern: namely, the status and nature of *materiality* vis-à-vis human life and experience; the nature of *time* and temporality (as unfolded through a world of things); and the nature and potential of *hope*, including in the face of historical conjunctures in which hope might seem to be increasingly dim.

The first of these, as laid out in the editors' introduction, concerns the question of materiality and its status vis-à-vis human order and action. This debate, now roiling the social sciences and humanities in general, has of course a long and distinguished STS pedigree. From classic work in actor-network theory and feminist technoscience to the more recent contributions of scholars like Barad (2007) and Orlikowski (2007), STS has played a central role in (re)introducing the social sciences to a world of things, showing how much of what we have traditionally claimed or attributed to "the social" has in fact deep and complex material roots and entanglements. Whether Latour's (1992) missing masses or Barad's (2007) agential realism this work has established a number of key propositions: the in-principle inseparability of social and material orders, the role of things (and the stories we tell about them) in anchoring and structuring social experience, and material engagement itself as a site of profound human value, meaning and attachment. These insights have subsequently come to inflect (infect?) contemporary scholarship more broadly, provoking what some have termed a materialist turn across the social sciences and humanities.

Growing scholarly concerns with repair both follow and extend this orientation. Repair studies (including those contained here) share the materialist ethos, eschewing idealisms or two-world accounts through which social and material things are held at arms-length, attending instead to how these worlds are connected and sustained through time. While agnostic in general on abstract theoretical debates around the agency of things, they accept and elaborate the role that things can play in the shaping of human value and experience. They are similarly attuned to the kinds of entanglements or configurations that recent materialist work has sought to foreground, and attentive to the silences and exclusions that such configurations may produce (including in the telling of our stories about them).

At the same time, repair studies may suggest holes, limits, and tensions within the materialist turn. One way of mapping these is to observe that this line of work may be as yet *insufficiently* materialized, and in particular has not yet paid adequate attention to certain routine operations and engagements by which the material is organized and held in place: thus, while the growing concern with materiality has cast fresh light on numerous aspects of the reciprocal shaping of human and material orders, it has paid less attention to the rich and specific relationships underpinning the longue durée relations that connect and sustain human and material orders in concrete times and places. Within technology studies, this shows up in the important but arguably privileged status assigned to "design" or "construction" as a moment through which values and ethics get inscribed into the built world. While recent attention to use and users within STS has helped to rebalance this story, it leaves out important and specific dimensions of work and skill that are badly named under the categories of either design or use. Repair helps to fill this hole, pointing to other kinds of material engagement that are no less complex and entangled, and which may play an outsized role in the sustenance and endurance (for better and for worse) of the sociomaterial orders so arrived at. If this helps speak to the *efficacy* of order, suggesting mechanisms of endurance and change that sidestep the various problems of determinism that stem from a two-world account, it also speaks to its *vibrancy* and often its precarity: a reminder that sociomaterial order, where it is to be found, is a hard-won and ongoing accomplishment.

The study of repair may offer similarly fresh insights into long-standing social science debates around time. As the editors rightly observe, the temporal standing of repair is at once multiple and ambiguous. On one hand, repair may be conceived as conservative and backward looking, oriented to returning broken objects and systems to prior states of function. On the other, repair may be viewed as forward looking and progressive, a site of *in situ* creativity and innovation that reworks the landscape of inherited situations to sustain and project value (and values) into the future. (The essays here of course provide rich examples of both). One response to this multiplicity of senses is to split them apart and name them differently: for example, in the useful typology offered by Sennett (2013), which breaks repair out into the distinct categories of restoration, remediation, and reconfiguration, each with a different temporal character and orientation. Running between these categories is the shared question of how exactly relationships between past and future are to be conceived and accomplished. This isn't always a question that's come naturally to social scientific ears, which have too often presumed stasis (or worse: unexplained teleologies) as the essential temporal character of the material world. In this context, the temporal ambiguity or multiplicity of repair may stand as feature rather than bug: an indication not of incoherence but of the complexity of the real-world operations that transgress (and thereby constitute) the shifting and ephemeral line between past and future states.

The last and most speculative theme broached here concerns repair and the problem of hope—a concept that has attracted at best mixed reviews within the canon of modern social theory. A long line of Marxist and other thinkers have been rightfully suspicious of the concept, seeing in hope a too-easy slide into forms of delusion and false consciousness that obscure the real-world operations of power that shape and constrain the human condition. Albert Camus (2005), for example, equates hope with resignation and defeat, arguing that of all the ills unleashed by Pandora, hope is the most dreadful. In the contemporary world, hope has served as engine and goldmine for modern marketers and demagogues alike, who have learned to deploy it in sophisticated, venal and sometimes destructive ways. As scholars like Lauren Berlant (2011) have noted, such deceptions may be not only false but *cruel*, an increasingly vain and poignant displacement of agency to the future even as the real conditions for personal and collective change grow fraught and dim. Under neoliberal sensibilities this cruelty is revisited upon the individual themselves as a kind of moral failing: "you had this freedom, why didn't you make more of it?"

But if hope can be thin and cruel, it can also be thick, grounding, and productive, a source of individual and collective agency rather than resignation. This sense underlies some of the more unconventional thinkers in the Marxist canon: for example, Ernst Bloch (1986 [1954]), who makes of hope an enduring principle that resists the predations of capital and fascism; or Walter Benjamin (1969), for whom hope amidst destruction and despair persists as an ever-present strain of messianic time within history. It has also been a defining feature and orientation of American pragmatism in both its classical (James and Dewey) and contemporary (e.g., Rorty) variants, underlying both the specific formulations of pragmatist thought and its commitment to programs of social change and reconstruction: so much so that recent interpreters have argued that "the characteristic attitude of the pragmatist is hope" (Koopman 2009: 17). These classic lineages have been joined by a recent resurgence in hope across a range of scholarly fields, from anthropology and sociology (where it has been deployed to explain empirical phenomena ranging from century-long waits for the restoration of ancestral homelands (Miyazaki 2004) to the zeitgeist of financial traders and other economic actors (Swedberg and Miyazaki 2016)) to philosophy (where scholars have begun to tackle hope as the neglected term in Kant's trio of great philosophical questions: "What can I know? What should I do? What may I hope?" (Chignell 2016). Still other philosophers—including several central to the lineage of STS—have argued for hope as a necessary value underpinning human responsibility and commitment to the animate and inanimate worlds around us (Zournasi and Serres 2002a, b).

What might repair have to do with this live and poignant debate? Here again, interesting questions arise as we attempt to think traditionally sociological or philosophical concepts into a material register. What *is* hope, as expressed through our material engagements with the world? Does the material world shape or participate in hope in any way—and if so, *which* ways? Is there anything inherently hopeful about acts of repair? If forms of hope practiced through repair can be quietist and conservative, can they also be critical and political? And where might hopes operated through repair intersect with *other* forms of relationality and attachment that have also attracted the periodic attention of repair and STS scholars: for example, reciprocity, love, and care?

Like most speculations and provocations, the points above remain partial, tentative, and contestable. In no way do I suggest that all repair workers operate upon an explicit basis and consciousness of hope (indeed, as the work contained here and elsewhere suggests, motivations for repair are multiple and diverse). Nor do all acts of maintenance and repair fix and connect time in as artful a way as suggested here. And maintenance and repair represent just one moment or aspect of the forms of material engagement that the social sciences are learning once again to consider. Nevertheless, the conjecture of this essay and volume remains. Taking repair seriously can help us towards more timely, materialized, and hopeful ways of thinking, making and fixing the worlds around us. The studies contained here are an excellent place to start.

References

Barad, Karen. 2007. *Meeting the Universe Halfway: Quantum Physics and the Entanglement of Matter and Meaning*. Raleigh, NC: Duke University Press.

Benjamin, Walter. 1969. *Illuminations: Essays and Reflections*. New York: Schocken Books.

Berlant, Lauren. 2011. *Cruel Optimism*. Raleigh, NC: Duke University Press.

Bloch, Ernst. 1986 [1954]. *The Principle of Hope*. Cambridge: MIT Press.

Bowker, Geoffrey C. 1994. *Science on the Run: Information Management and Industrial Geophysics at Schlumberger, 1920–1940*. Cambridge: MIT Press.

Bowker, Geoffrey C., and Susan Leigh Star. 1999. *Sorting Things Out: Classification and Its Consequences*. Cambridge: MIT Press.

Camus, Albert. 2005. *Summer in Algiers*. London: Penguin.

Chignell, Andrew. 2016. *What May I Hope?* New York: Routledge.

Edwards, Paul N., Steven J. Jackson, Geoffrey C. Bowker, and Corey P. Knobel. 2007. *Understanding Infrastructure: Dynamics, Tensions, and Design*. Report of the National Science Foundation Workshop, "History and Theory of Infrastructure: Lessons for New Scientific Cyberinfrastructures." Ann Arbor: DeepBlue. http://hdl.handle.net/2027.42/49353.

Edwards, Paul N., Steven J. Jackson, Geoffrey C. Bowker, and Robin Williams. 2009. An agenda for infrastructure studies. *Journal of the Association for Information Systems* 10: 364–374.

Garfinkel, Harold. 1967. *Studies in Ethnomethodology*. New York: Prentice-Hall.

Harman, Graham. 2002. *Tool-Being: Heidegger and the Metaphysics of Objects*. New York: Open Court.

Henke, Christopher R. 2000. The mechanics of workplace order: Toward a sociology of repair. *Berkeley Journal of Sociology* 44: 55–81.

Jackson, Steven J. 2014. Rethinking repair. In *Media Technologies: Essays on Communication, Materiality, and Society*, ed. Tarleton Gillespie, Pablo J. Boczkowski, and Kirsten A. Foot, 221–239. Cambridge: MIT Press.

James, William. 1904. A world of pure experience. *Journal of Philosophy, Psychology, and Scientific Methods* 1 (20): 533–543.

Koopman, Colin. 2009. *Pragmatism as Transition: Historicity and Hope in James, Dewey, and Rorty*. New York: Columbia University Press.

Latour, Bruno. 1992. Where are the missing masses? Sociology of a few mundane artifacts. In *Shaping Technology, Building Society: Studies in Sociotechnical Change*, ed. Wiebe E. Bijker and John Law, 225–258. Cambridge: MIT Press.

Melville, Herman. 1851. *Moby Dick: Or, The Whale*. New York: Harper and Brothers.

Miyazaki, Hirokazu. 2004. *The Method of Hope: Anthropology, Philosophy, and Fijian Knowledge*. Stanford: Stanford University Press.

Orlikowski, Wanda. 2007. Sociomateriality: Exploring technology at work. *Organization Studies* 28: 1435–1448.

Sennett, Richard. 2013. *Together: The Rituals, Pleasure and Politics of Cooperation*. New Haven, CT: Yale University Press.

Swedburg, Richard, and Hirokazu Miyazaki (eds.). 2016. *The Economy of Hope*. Philadelphia: University of Pennsylvania Press.

Zournazi, Mary and Isabelle Stengers. 2002. A 'cosmo-politics': Risk, hope, change—With Isabelle Stengers. In *Hope: New Philosophies for Change*, ed. Mary Zournazi, 244–272. New York: Routledge.

Zournazi, Mary, and Michel Serres. 2002. The art of living—With Michel Serres. In *Hope: New Philosophies for Change*, ed. Mary Zournazi, 192–208. New York: Routledge.

Index

A

Actor-network theory (ANT) 3, 17, 20, 64, 71, 83, 92

Agency 3, 5, 32, 134, 156, 164, 170, 179, 259, 274, 322, 326, 339, 342, 344

Assemblage(s) vi, vii, 17, 46, 141, 142, 162, 164–167, 170, 173, 175–177, 179–181, 339, 340

B

Bricolage v, 7, 52, 172, 173, 199, 261, 340

C

Collective(s) 6, 19, 21, 37, 56, 221, 234, 235, 243, 249, 313–317, 319–328, 330–332, 344

Consumer(s) 283, 285, 287, 290, 313

Conversational 11, 113, 257, 259

Conversation(s) 14, 15, 20–22, 72, 73, 78, 140, 208, 228, 257, 325, 337

D

Dant, Tim v, 4, 17, 19, 21, 22, 148, 165, 169, 172, 188, 195, 228, 241, 245, 249, 250, 259, 261, 266, 293, 304, 307, 318, 338, 339

Dewey, John 5, 6, 8, 12–14, 18, 35–37, 40, 45, 47, 56, 189, 197, 198, 215, 345

Diagnosis 12, 13, 22, 69, 139, 148, 149, 236, 237, 246, 247, 314, 327

I. Strebel et al. (eds.), *Repair Work Ethnographies*,
https://doi.org/10.1007/978-981-13-2110-8

Discursive 18, 70, 99, 258, 263, 265, 267, 272, 273, 275, 276
Do-it-yourself 228, 249, 283, 284, 318

E

Empowerment 322, 326
Ethnographer(s) 9, 35, 71–73, 100, 263, 275–277
Ethnography viii, 7, 11, 13, 15–19, 64, 71, 80, 83, 90, 91, 165, 179, 181, 221, 222, 249, 258, 265, 274–277
Ethnomethodology 6, 12–15, 22, 46, 257

G

Garfinkel, Harold 5, 6, 8, 12–14, 35, 44, 56, 69, 91, 93, 257, 340
Gender 19, 140, 264, 265, 274, 316–319, 326, 330–332, 340

H

Harper, Douglas v, viii, 7–11, 45, 91, 240, 261, 263, 266, 267, 270
Heidegger, Martin 4, 5, 8, 285, 286
Henke, Christopher R. v, 7, 12–14, 18, 19, 43, 46, 52, 67, 133, 172, 188, 190, 195, 200, 256, 257, 260, 262, 264, 266, 268–270, 272–274, 276, 284, 339
Heuristic(s) 2, 4–6, 8, 18

I

Improvisation(s) v, 15, 32, 48, 50–56, 130, 134, 165, 172, 173, 200, 233
Infrastructure(s) 1, 18, 20, 22, 81, 89, 92, 110, 117, 120, 121, 129, 130, 132, 136, 164, 165, 180, 181, 188, 189, 209, 211, 212, 214, 224, 256–260, 262, 263, 266–270, 272–276, 339–341
Innovation(s) vi, 2, 17, 64, 187, 188, 190, 197, 199, 200, 202, 211, 212, 241, 249, 343
Inquiry 2, 5–7, 12–18, 21, 32, 35–37, 40, 41, 44–47, 56, 66, 81, 97–99, 187–190, 194, 197–199, 201, 203, 207, 210, 212–214, 339, 340
Invisibility 165, 173, 176, 182, 222, 260, 340

L

Latour, Bruno 3, 62, 78, 164, 188, 196, 258, 260, 342

M

Materiality 2, 3, 19, 46, 79, 92, 119, 200, 212, 257–260, 263, 267, 274–276, 342, 343

O

Orr, Julian E. v, 7, 11–13, 52, 54, 100, 120, 131, 133, 134, 150, 152, 155, 157, 169, 172, 234, 243, 260, 265, 276

P

Phenomenology 181
Pinch, Trevor viii, 2, 3, 14, 46, 52, 54, 118, 216
Politics 2, 15, 18, 20, 81, 256, 258, 273, 276, 318, 319, 338
Pragmatism 6, 15, 46, 213, 345

R

Routine(s) 14, 15, 17, 32, 33, 35, 41, 42, 45, 48, 53, 54, 65, 69, 70, 91, 93, 111, 116, 123, 173, 199, 247, 339, 343

S

Safety 33, 38–40, 43, 55, 173, 177, 178, 195, 216, 250, 309
Science and technology studies (STS) vi, 2, 3, 13, 14, 20, 71, 79, 119–121, 257, 341–343, 345

T

Temporality 113, 293, 298, 342
Transformation(s) 5, 7, 80, 82, 104, 116, 196, 207, 210, 270, 272, 273, 292

V

Video(s) 3, 15, 16, 90–92, 94, 96, 98, 101, 110, 112, 115, 117, 121, 123, 287–290, 293–307, 309, 311
Visibility 17, 176, 181, 182, 260, 329, 340

The manufacturer's authorised representative in the EU is Springer Nature Customer Service Centre GmbH, Europaplatz 3, 69115 Heidelberg, Germany. If you have any concerns regarding our products, please contact ProductSafety@springernature.com

Printed and bound by CPI Group (UK) Ltd, Croydon, CR0 4YY
25/09/2026
02222894-0005